Recognition of Health Hazards in Industry

RECOGNITION OF HEALTH HAZARDS IN INDUSTRY
A Review of Materials and Processes

WILLIAM A. BURGESS
Associate Professor of Occupational Health Engineering
Harvard School of Public Health
Boston, Massachusetts

A WILEY-INTERSCIENCE PUBLICATION

JOHN WILEY & SONS   New York • Chichester • Brisbane • Toronto

Library of Congress Cataloging in Publication Data:

Burgess, William A., 1924-
  Recognition of health hazards in industry.

  "A Wiley-Interscience publication."
  Includes index.
  1. Industrial hygiene. I. Title. [DNLM:
1. Occupational medicine. WA 400 B955r]
HD7261.B94   620.8'5   81-2132
ISBN 0-471-06339-8   AACR2

Printed in the United States of America

10 9 8 7 6 5 4 3 2 1

*To occupational health professionals
and the cooperative spirit
in which they perform their roles*

PREFACE

This book is designed to introduce the reader to industrial operations and how these operations can affect the health of workers.

It is hoped that this volume will be of value to the student pursuing an undergraduate or graduate degree in occupational health, the physician who seeks to relate medical symptoms to job exposures, the plant engineer, and management and union health and safety representatives who have the responsibility for assuring that industrial processes do not present health hazards to workers. This volume should also serve as a practical field reference for industrial hygienists in insurance, consulting, and compliance activities who encounter a wide range of industrial operations.

This book has been written with this diverse audience in mind. It is not encyclopedic in terms of the operations covered or in terms of the literature citations. One or two major references (denoted by an asterisk) are recommended on each unit operation in Chapter 2 for the individual who desires more information.

Manufacturing processes when first viewed, frequently present a barrier to the occupational health specialist because of the variety of equipment and materials in use and the associated terminology. Until the 1970s this was not a great problem since persons entering the health and safety field frequently had engineering or chemistry training and many had industrial experience. Today, however, it is common for many people to enter the field directly after having completed their academic training. For those in this category, general descriptions of processes and related health hazards are needed.

The material in this book is presented in three chapters. Chapter 1 covers the general format of a plant survey and the information that should be obtained. In Chapter 2, the common unit operations encountered in many different industrial settings are described in detail with appropriate literature citations. Chapter 3 contains a brief review of selected industrial production facilities. The industries chosen are major production facilities employing large numbers of workers and representing varying hazard potentials.

In each case materials and equipment are described in detail using nomenclature appropriate to the industry. The physical form and origin of the air contaminants are identified as well as the physical stresses encountered in the process or industry. The controls one would expect to see on operations are noted so that one can gauge the conformity of a given plant with good engineering practice. Where air sampling data on similar operations are available, they are presented.

WILLIAM A. BURGESS

Boston, Massachusetts

June 1981

ACKNOWLEDGMENTS

This book has its origins in a chapter on potential exposures in industry published in the revised edition of *Patty's Industrial Hygiene and Toxicology*. The discussion of glassmaking was adapted from material written by the late J. Dunn for the Original Patty Volume. The section on industrial radiography has been prepared by John J. Munro III. Flow charts prepared by the Environmental Protection Agency have been used extensively in Chapter 3. Other useful information has been obtained from numerous publications of the National Institute for Occupational Safety and Health and the American Conference of Governmental Industrial Hygienists.

As a general practitioner of industrial hygiene, I have surveyed most of the operations discussed in this volume. Additional perspective and helpful criticism have been provided by Darrel E. Anderson, James C. Barrett, Thomas B. Bonney, James E. Hickey, Leon D. Horowitz, Marvin E. Kennebeck, Arthur Lange, Mars Y. Longley, James S. McKarns, Franklin E. Mirer, Gary E. Mosher, Ernest Neukuckatz, Leonard D. Pagnotto, Allen Storm, Russell W. Van Houten, Fred S. Venable, Donald L. Webster, James L. Weeks, and John D. Yoder.

Several industries, industrial associations, and unions have contributed material, and their contributions have been acknowledged in the text. Colleagues in industry and faculty and students at the Harvard School of Public Health have been helpful in reviewing manufacturing operations and sharing their experiences on health hazards in the workplace.

I am grateful to Kenneth P. Martin for assistance in preparing the figures and to Jacquelynne Bigelow, Esther Lumbert, Susan Sheridan, and Claire Wasserboehr who typed early drafts and the final copy of the text.

I owe special thanks to Robert Treitman for reading the entire manuscript and offering many useful suggestions.

W. A. B.

CONTENTS

Recognition of Health Hazards in Industry

INTRODUCTION 1

The first step to be taken in a preventive program in occupational health is the identification or recognition of potential health hazards. The most effective technique to accomplish that task is the preliminary plant survey.

Mastery of the preliminary survey of the workplace is based on thorough knowledge of industrial materials and processes. Without this knowledge it is difficult for the surveyor to identify those industrial processes that have the potential to cause occupational disease. This volume is designed to assist occupational health personnel in preparing for a preliminary survey by describing industrial processes and their impact on the workplace environment. Chapter 2 is a review of unit operations and Chapter 3 introduces a description of a selected number of industries. The goal of this volume is to provide the reader with understanding of the industrial operations including familiarity with in-plant terminology, the identity and sources of air contaminants that may be released to the workplace, and the physical stresses introduced by such processes.

1.1 MATERIALS

In Chapter 2 information is presented on materials including chemicals, plastics, and metals in use in common industrial processes. A system of material specifications including composition and physical and chemical properties has evolved in the United States; without such specifications American industry could not function. These specifications for industrial materials are useful in evaluating worker exposure and are frequently cited in the text. Detailed material data and references to other sources of information on metal specifications are presented in Appendix A.

How does one determine what materials are used in the plant? Prior to the visit to the facility the surveyor can obtain preliminary information by referring to the various engineering and chemistry texts listed in Appendix B that contain process flow diagrams and general descriptions of materials. Although some manufacturing facilities do have proprietary materials and processes, most of the standard industrial goods are produced in a remarkably similar fashion.

In large manufacturing plants the detailed process information maintained by the company will include lists with raw material specifications and approved vendors. These data are usually not available on the production floor but can be obtained from the product engineer, material control laboratory personnel, or the purchasing agent. In plants with proprietary processes, this information may be available from senior management persons and its release must usually be negotiated.

A raw material list may not be available in the small plant manufacturing a conventional product. Standardization of materials in these plants is frequently not necessary since variations in composition of materials and chemicals have little effect on the product. A material list may be generated by direct observation at the work site or by conducting an inventory of the material in the receiving dock or warehouse.

Most of the chemicals used in industry are technical grade; that is, they are not pure chemicals but contain certain impurities. In a sophisticated process these impurities may affect the yield and quality of the product. In this case the manufacturer will have detailed material specifications, and routine analysis of incoming materials will be done by the material control laboratory. These data may be helpful in the plant survey. If the impurity is present as a fraction of a percent and it is not a highly toxic material, it probably will not present a health hazard. If it is a highly volatile and toxic chemical, it must be considered a potential hazard. Frequently one encounters chemicals where the concentration of impurities may be as high as 20%. Such a case warrants detailed review. The issue of impurities is further complicated when their identity may be unknown even to the manufacturer of the material.

Information on the amount of raw material consumed in a manufacturing process is of value in considering the relative risk to the worker. If only a few kilograms per month are used, it may not warrant follow-up. If large quantities are used, one would think that a material balance could be calculated and the quantity of chemical released to the environment could be estimated. The ventilation rates could then be determined, and the air concentration could be calculated. This is rarely successful. In general, use rates are only valuable in making qualitative assessments of exposure; one must rely on air sampling during the evaluation step to determine exposure levels.

1.2 PROCESSES

Large volume standard chemicals and products are manufactured with rather similar processes and materials. The best description of the manufacturing process is obtained from the plant and most facilities will have such information, usually in the form of a flow chart. Engineering and chemistry texts cited in Appendix B provide similar data. The process nomenclature may differ between plants and the investigator should take care to clarify these differences before completing the survey. Plant personnel will not place much faith in occupational health reports that do not reflect knowledge of the plant and the associated terminology.

Many industrial associations provide educational information on the standard operations in their industry. Frequently, excellent flow charts are available and the larger industrial associations have ongoing occupational health committees that may provide reports on industrial activities. A list of associations providing such information is presented in Appendix B.

Other sources of information on industrial processes are air pollution publications that provide detailed pollution emission rates from industrial operations. These reports are useful in occupational health studies since the flow diagrams identify the contaminant release points, the nature of the contaminant, and the quantity released.

1.3 EXPOSURE PATTERN

Once the operation has been defined and the contaminant release points identified, the time pattern during which the worker is exposed must be established. Since approximate data are usually adequate, one should not request an industrial engineering time study.

On selected operations, it may be important to obtain accurate exposure times as a first step in identifying the potential hazards. For example, if a welder is performing a job that requires a long setup time he may only be welding 5–10% of the shift time. However, in another operation a welder may be working on large metal parts, and arc time may be as high as 60–70% of the shift. In this case, arc time can be estimated by viewing the operation. If an accurate assessment is needed, an elapsed time device can be placed on the welding equipment. In other processes, utilization time on equipment and the quantity of material used during a shift provide an index of exposure time that may be helpful in assessing the hazard.

1.4 CONTROLS

There are conventional controls one expects to see in place on certain industrial operations. The plant's failure to utilize these controls may suggest a lack of knowledge of or indifference to the control of health hazards. The common controls utilized in industry to minimize risk to employees are engineering controls, work practices, and personal protective equipment. The application of these controls to specific processes is covered in Chapters 2 and 3.

1.4.1 Engineering

One of the most effective engineering controls is ventilation; the principal source of ventilation design data is the American Conference on Governmental Industrial Hygienists (ACGIH) Industrial Ventilation Manual, and this source is referenced frequently in the text (1). The designs in the Ventilation Manual represent best available engineering practice. Although these designs have been shown to control exposure to acceptable levels in industry, the capture efficiency of hoods under plant conditions has not been evaluated. The design plates provide a description of the hood geometry, the airflow rate, duct transport velocity, and entry loss.

The desirable procedure in implementing ventilation controls is that an engineering design is proposed, installation is made based on the design, and the completed ventilation system is evaluated to ensure it meets the minimum design specifications and provides effective control of the air contaminant. One should not concede *a priori* that control is achieved just because a system has been installed and looks similar to the recommended design.

A number of simple field observations can be made to assess the value of the local exhaust ventilation system. If the system has been evaluated, there is usually some evidence—either visible holes in the duct indicating that duct velocities have been measured or a notation of velocity or exhaust volume written on the hood. Plant engineering may have the system on their preventive maintenance schedule and records of performance may be available for comparison with recommended standards. The occupational health specialist inspecting ventilation systems can often determine their adequacy by making simple checks. Examine the duct to see if it is intact or if it has been damaged. If the system handles particulates, knock on the duct. A hollow sound indicates that transport velocities are adequate for conveying particulates. A dull thud suggests that dust is settling out in the duct. If a new branch duct has been added to the main of an old system, the performance of the original system may be impaired. One should follow the duct run and determine if a fan is installed in the system. If a fan is installed, determine if it is operating in the correct direction. Centrifugal blowers will move 30–40% of their rated capacity while running backwards. A smoke tracer released at the hood capture point can be used to establish the need for further evaluation of the system.

Engineering controls may be an integral part of the process equipment or they may require action by the worker. One should ensure that installed controls are not by-passed by the operator or the supervisor in the interest of comfort or increased productivity. As an example, degreasers are frequently equipped with a low speed hoist to minimize solvent loss to the room. If such equipment is available, check to see that it is routinely used. Dipping parts by hand may be faster but the resulting solvent "pull out" is reflected in high solvent concentration.

1.4.2 Work Practices

Work practices should only be considered an integral part of a control program if they have been documented, all employees have been informed of the work practices, and they are a part of the training program for new employees. Work practices may develop at the plant over a period of years or they may be introduced as a result of an Occupational Safety and Health Administration (OSHA) standard. The workers may be a valuable resource in developing procedures to minimize exposure.

In evaluating the effectiveness of published work practices observe the workers to determine if they all use the same procedures in completing the work. If each worker does his "own thing," the exposure patterns will vary widely. At a bagging operation in a talc plant, exposure concentrations varied by an order of magnitude depending on the individual work practice of the operators. In a rayon plant, individual differences in work practices permitted some employees to continue work while others had to be removed from the carbon disulfide exposure.

In this volume, specific work practices are presented when available in the literature. In general, however, the importance of work practices in the control of exposures has not been recognized and few comprehensive statements have been published.

1.4.3 Personal Protective Equipment

With the advent of OSHA and the increased emphasis on engineering control, it was anticipated that the utilization of personal protective equipment including respirators would decrease during the 1970s. In fact, the use of such personal protective equipment expanded.

The application of respiratory protection is described in regulatory documents and in an application guide for respirators based on the National Institute for Occupational Safety and Health (NIOSH)/OSHA Standard Completion Program (2). The use of other protective equipment is less precisely defined but useful guidelines are available. In this book comments are made on critical applications of personal protective equipment in processes such as welding; however, routine applications are not mentioned since they are thoroughly covered in other publications.

In reviewing the application of respirators in the plant, one should review not only the appropriateness of the respirator issued to the worker, but the total respiratory protection program. The most effective procedure to ensure proper utilization of equipment for a given process is to specify its use as a part of manufacturing process instructions. One can evaluate qualitatively the effectiveness of the personal protective equipment program by noting whether the equipment is treated and stored like a valuable tool. If this is done, a good program is probably in place. If the equipment is hanging from a valve stem or thrown in the bottom of a locker, it is probably not an effective program.

1.5   EVIDENCE OF EXPOSURE

It is frequently stated that an inexperienced person can use his or her vision and sense of smell to identify and estimate the concentration of a contaminant. This is an overstatement. One can just see a 25 μm particle under ideal lighting conditions. In a plant one cannot, therefore, see respirable dust of less than 10 μm unless a dense cloud of dust, mist, or fume is generated. Small particles can be identified by light scatter using a window or a portable lamp as the light source. This simple technique is useful in engineering control, but one cannot use it to establish dust concentrations.

The presence of gases and vapors may be identified by odor or irritation, and sources of information on odor threshold data are noted in Appendix B. These sources should be part of the desk references of the occupational health specialist.

1.6   INFORMATION SOURCES

The OSHA Permissible Exposure Limits (PELs) (3), NIOSH recommended standards (4), and ACGIH Threshold Limit Values (TLVs) (5) are not included in this book nor are the toxic effects of materials routinely noted. The standards are continuously revised and can be obtained easily from other sources. This is not a volume on the toxicity of industrial materials; other sources are available on the effects of exposure to contaminants (6, 7, 8).

The physical forms of air contaminants released from each process are noted in the text and defined in Appendix C. This information is necessary in assessing exposure and evaluating the appropriateness of air cleaning equipment.

1.7   FIELD NOTES

Although this volume is concerned only with recognition of potential health hazards preparatory to later evaluation and control, one should not discount the need for reporting the field observations in this phase of the work. A flow sheet and a diagram of the plant identifying the processes, workers exposed, and control techniques are necessary for proper reporting. The nomenclature used in the plant must be referenced to standard industry terminology. The raw materials and input chemicals used in the process, the intermediate chemical by-products, and the final product must be identified. The release points must be noted on the diagram of the process. The type of ventilation, that is dilution or local exhaust ventilation as defined in the ACGIH Ventilation Manual, must be noted with simple line sketches. Some assessment of the quality of the ventilation should be included in the field notes.

REFERENCES

1 Committee on Industrial Ventilation, American Conference of Governmental Industrial Hygienists, *Industrial Ventilation: A Manual of Recommended Practice,* 16th ed., ACGIH, Lansing MI, 1980.

2 "Pocket Guide to Chemical Hazards," U.S. Department of Health, Education and Welfare, Publication No. (NIOSH) 78-210, Cincinnati, OH, 1978.

3 Federal Occupational Safety and Health Standards (29 CFR 1910, Subpart Z).

4 "Criteria for a Recommended Standard—Occupational Exposure to _____" U.S. Department of Health, Education and Welfare, NIOSH, Cincinnati, OH, 19XX.

5 American Conference of Governmental Industrial Hygienists, *TLVs—Threshold Limit Values for Chemical Substances in the Workroom Environment with Intended Changes,* American Conference of Governmental Industrial Hygienists, P.O. Box 1937, Cincinnati, OH 45201, published annually.

6 N. H. Proctor and J. P. Hughes, *Chemical Hazards of the Workplace,* Lippincott, Philadelphia, PA, 1978.

7 Marcus M. Key, Austin F. Henschel, Jack Butler, Robert N. Ligo, and Irving R. Tabershaw, Eds., *Occupational Diseases: A Guide to their Recognition,* rev. ed. U.S. Department of Health, Education and Welfare, Washington, D.C., June 1977.

8 G. D. Clayton and F. E. Clayton, *Patty's Industrial Hygiene and Toxicology,* Vol. 2, 3rd ed., Wiley-Interscience, New York, 1981.

INDUSTRIAL UNIT OPERATIONS

2

2.1 ABRASIVE BLASTING

Abrasive blasting is encountered in a number of occupational settings including bridge and building construction, shipbuilding and repair, founding, and metal finishing. The process is used in heavy industry to remove surface coatings, scale, rust, or fused sand in preparation for subsequent finishing operations. No other method is as effective and economical for this purpose. Abrasive blasting is also used in intermediate finishing operations to remove flashing, tooling marks, or burns from cast, welded, or machined fabrications and to provide a matte finish to metal parts. In addition, chilled-iron and steel shot abrasive can be used to peen the metal surface to improve resistance to fatigue stress, reduce surface porosity, and increase the wearability of parts.

Various abrasives are used in abrasive blasting installations. Ground corn cobs, nut shells, and glass beads and spheres are used for light cleaning while removing little metal. The common heavy duty abrasives are silica sand, aluminum oxide (Al_2O_3), silicon carbide (SiC), blast furnace slag, iron and steel shot and grit, and aluminum shot. The frequency of use of these various materials in one segment of the industry is shown in Table 2.1-1 (1).

Three major types of blasting equipment are used in the industry. Air blasting equipment either uses the abrasive in a pressurized air tank that is fed to the blasting hose or compressed air aspirates the abrasive through a flexible hose to a nozzle where it is directed to the workpiece at high velocity. A high speed impeller is used in centrifugal units to project the abrasive at the workpiece. The wet blast technique utilizes a suspension of abrasive in water or, occasionally, petroleum distillate that is directed at the workpiece from a nozzle using water pressure or air and water pressure.

Table 2.1-1 *Abrasives*

Abrasive	Facilities Using Abrasive (%)
Sand	44.7
Steel shot	16.7
Steel grit	9.7
Aluminum oxide	9.3
Flint/garnet	7.0
Glass beads	4.6
Silicon carbide	3.5
Slag	3.1
Organic	1.1

Source. Reference 1.

2.1.1 Application and Hazards

In open air blasting in construction and shipyard applications, general area contamination occurs unless the work area is enclosed with a temporary barrier. The blasting crew, including the operator, "pot man," and cleanup personnel, may be exposed to high dust concentrations depending on the existing wind conditions. Open air work is difficult to control and has resulted in silicosis in blasters and nearby workers after brief exposure periods.

Vacuum blasting is a relatively new technique that is effective in outdoor abrasive blasting operations where one cannot provide an exhausted enclosure for dust control. The abrasive nozzle is coupled with a low volume–high velocity exhaust system that provides pickup of spent abrasive and workpiece debris directly at the site of blasting. No data have been published in the open literature on the collection efficiency of this system.

In the metal working industry a variety of exhausted enclosures (Figures 2.1-1–2.1-4) have evolved including cabinets, blasting rooms, automatic barrel, rotating table, and tunnel units (2). In most cases these industrial units have integral local exhaust ventilation systems.

The two obvious health hazards one must consider in reviewing abrasive blasting operations are inhalation of airborne contaminants and noise. In evaluating the exposure of the worker to airborne dust, one must identify the base metal being blasted, the nature of the surface coating or contamination being removed, and the abrasive in use and its contamination from previous operations.

In the United States the widespread use of sand containing high concentrations of crystalline quartz continues to present a major hazard to

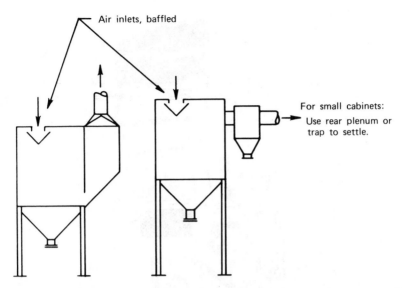

Figure 2.1-1 Abrasive blasting cabinet. Minimum exhaust volume is based on 20 air changes per minute. (From Reference 2)

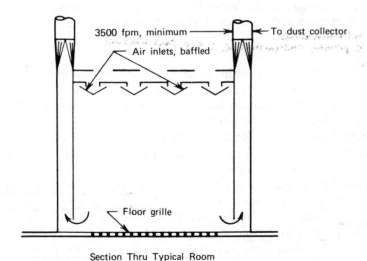

Figure 2.1-2 Abrasive blasting room. Minimum downdraft ventilation of 80 feet per min. over projected floor area. (From Reference 2)

Figure 2.1-3 Abrasive blasting barrel. Minimum exhaust volume of 500 feet per min. through all openings with door closed. (From Reference 2)

Noise

Dust
- *from surface contamination*
- *from abrasive ± previous contamination : esp sand blasting*
- *from base metal*

workers. A number of European nations have forbidden the use of sand for this application, but this effective control measure has not been adopted in the United States. The other organic and inorganic abrasives are classed as nuisance particulates by ACGIH.

In most cases the base metal being blasted is iron or steel and the resulting exposure to iron dust will present a limited hazard. If metal alloys containing such materials as nickel, manganese, lead, or chromium are worked, the hazard should be evaluated by air sampling. The surface contamination on the workpiece may also present a major hazard. In the foundry it may be fused silica sand; if one is involved in ship repair, the surface coating may be a lead-based paint or an organic mercury fungicidal coating. In addition to direct exposure to the dust from the surface coating, one must guard against concentrating the coating contaminant in the abrasive recycle system. Abrasives used to remove lead paint from a bridge structure may contain up to 1% lead by weight.

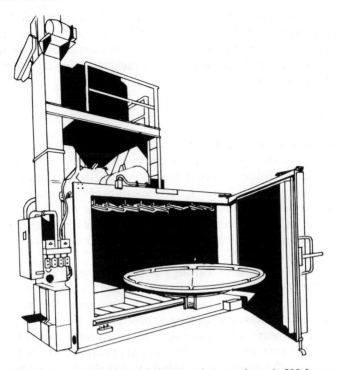

Figure 2.1-4 Rotary blasting table. Minimum exhaust volume is 500 feet per minute through all openings with door closed. (From Reference 2)

Table 2.1-2 *Abrasive Blasting Exposure Data*

Industry	No. of Samples	Respirable Mass Dust Concentrations, (mg/m³)		Average Respirator Protection Factors
		Min.	Max.	
Monument	13	0.4	3.7	5.3
Shipyard	16	2.5	73.6	235
Painting/sandblast- ing contractors	36	1.3	268.0	60
Primary metals				
Air blasting	26	1.3	142.9	128
Airless blasting	21	—	84.0	2.5

Source. Reference 1.

The range of air concentrations of respirable dust for a variety of industries noted in one survey is shown in Table 2.1-2 (1). These data are grab samples taken with a direct-reading instrument. Little data are available on time weighted average concentrations for these operations.

When blasting units are used in open air or in blasting rooms the operators are exposed to noise levels in the range of 85 to 118 dBA. The noise exposure at the operator's location at cabinets or automatic rooms ranges from 83 to 110 dBA. In a survey of 22 abrasive blasting facilities, 17 exceeded the present OSHA standard of 90 dBA (3).

2.1.2 Control

The ventilation requirements for abrasive blasting enclosures have evolved over several decades, and effective design criteria are now available. The minimum exhaust volumes, based on seals and curtains in new condition, are shown in Figures 2.1-1–2.1-4 (2, 4, 5). The nature of the operation necessitates rugged enclosures and easily maintained curtains and seals to eliminate outward dust leakage. Systems to be used for organic abrasives require special design features because of the fire and explosion hazard and operators should not be permitted to work inside such enclosures.

Dust control on abrasive blasting equipment depends to a large degree on the integrity of the enclosure. All units should be inspected periodically, including baffle plates at air inlets, gaskets around doors and windows, gloves and sleeves on cabinets, star gaskets at hose inlets, and the major structural seams of the enclosure. The dust collection system should also be inspected to insure that fines are removed and the necessary periodic maintenance is carried out. If possible, the operation in walk-in rooms should be set up so that the direct blast of the nozzle does not hit the door.

When the operator is directly exposed to the blasting operation as in open air or blasting room operations, he must be provided with a NIOSH-approved Type CE abrasive blasting hood or helmet. The intake for the air compressor providing the respirable air supply should be located in an area free from air contamination and the quality of the air delivered to the respirator must be checked periodically. The respirator must be used in the context of a full respirator program, and rigorous maintenance of the equipment is essential. Serious health hazards may result from inadequate abrasive blasting respirator programs (1). To evaluate accurately the protection from such equipment, air samples should be taken inside the helmet.

REFERENCES

1   A. Blair, "Abrasive Blasting Respiratory Protective Practices," U. S. Department of Health, Education and Welfare Publication No. (NIOSH) 74-104, Cincinnati, OH, 1974.

2   J. H. Hagopian and E. K. Bastress, "Recommended Industrial Ventilation Guidelines," U. S. Department of Health, Education and Welfare NIOSH Research Contract Report No. CDC-99-74-33, Cincinnati, OH, 1976.

*3   J. L. Goodier, E. Boudreau, G. Coletta, and R. Lucas, "Industrial Health and Safety Criteria for Abrasive Blasting Cleaning Operations," U. S. Department of Health, Education and Welfare Publication No. (NIOSH) 75-122, Cincinnati, OH, 1975.

4 Committee on Industrial Ventilation, American Conference of Governmental Industrial Hygienists, *Industrial Ventilation: A Manual of Recommended Practice*, 16th ed., ACGIH, Lansing, MI, 1980.

5 American National Standards Institute, "Ventilation and Safe Practices of Abrasive Blasting Operations," ANSI Z9.4—1979, ANSI, New York, 1979.

2.2   ACID AND ALKALI CLEANING OF METALS

After the removal of major soils, oils, and grease by degreasing, the metal parts are often treated with acid and alkaline baths to condition the parts for electroplating or other finishes. The principal hazards in this series of operations include exposure to acid and alkaline mist released because of heating, air agitation, gassing from electrolytic operation, or cross contamination between tanks.

2.2.1   Acid Pickling and Bright Dip

Pickling is a term describing the treatment of metals with acids and is apparently derived from the early practice of cleaning metal parts by dipping them in vinegar. Its principal application is for the removal of scale, rust, corrosion products, hard water scale, and light oxide coatings prior to surface finishing (1, 2, 3).

Scale and rust are commonly removed from low and medium alloy steels using a nonelectrolytic immersion bath of 5–15% sulfuric acid at a temperature of 60–82°C (140–180°F) or a 10–25% hydrochloric acid bath at room temperature. Since sulfuric and hydrochloric acids frequently cause pitting, phosphoric acid may be used to remove hard scale and water scale. When steel parts are pickled, a residue or smut is frequently left on the surface. This smut can be removed by an anodic electropickle, alkaline electrocleaning, or ultrasonic cleaning with an alkaline cleaning solution (3).

In order to remove the oxide film from stainless steels that have been heat treated, a two-stage pickling process is frequently used. For alloys containing greater than 17% chromium, the parts are first immersed in a 10% sulfuric acid bath operating at 66–82°C (150–180°F) to soften the scale, and then a bath of 25% hydrochloric, 10% nitric, and 1½% hydrofluoric acids operating at room temperature to remove the scale. For alloys containing less than 17% chromium one may use an immersion bath of 20% nitric and 3% hydrofluoric acid at 21–54°C (70–130°F) or a bath of 10% ferric sulfate and 2% hydrofluoric at 49–60°C (120–140°F). If the scale is not removed with this acid pickle, one must use an alkali scaling bath. The efficiency of these pickling operations can be improved by operating at higher temperatures, ultrasonically, or electrolytically. After pickling of stainless steels, an operation called passivation is conducted in nitric acid to provide corrosion resistance.

Pickling operations on nonferrous metals such as aluminum, magnesium, zinc, and lead can take several forms. Aluminum parts do not require acid

Table 2.2-1 *Contaminants Released by Pickling Operations*

Type	Component of Bath That May Be Released to Atmosphere	Physical and Chemical Nature of Major Atmospheric Contaminant
Aluminum	Nitric acid	Nitrogen oxide gases
Aluminum	Chromic, sulfuric acids	Acid mists
Aluminum	Sodium hydroxide	Alkaline mist
Cast iron	Hydrofluoric-nitric acids	Hydrogen fluoride–nitrogen oxide gases
Copper	Sulfuric acid	Acid mist, steam
Copper	None	None
Duraluminum	Sodium fluoride, sulfuric acid	Hydrogen fluoride gas, acid mist
Inconel	Nitric, hydrofluoric acids	Nitrogen oxide, HF gases, steam
Inconel	Sulfuric acid	Sulfuric acid mist, steam
Iron and steel	Hydrochloric acid	Hydrogen chloride gas
Iron and steel	Sulfuric acid	Sulfuric acid mist, steam
Magnesium	Chromic-sulfuric, nitric acids	Nitrogen oxide gases, acid mist, steam
Monel and nickel	Hydrochloric acid	Hydrogen chloride gas, steam
Monel and nickel	Sulfuric acid	Sulfuric acid mist, steam
Nickel silver	Sulfuric acid	Acid mist, steam
Silver	Sodium cyanide	Cyanide mist, steam
Stainless steel	Nitric, hydrofluoric acids	Nitrogen oxide, hydrogen fluoride gases
Stainless steel	Hydrochloric acid	Hydrogen chloride gas
Stainless steel	Sulfuric acid	Sulfuric acid mist, steam
Stainless steel		
Immunization	Nitric acid	Nitrogen oxide gases
Passivation	Nitric acid	Nitrogen oxide gases

Source.   Reference 4.

pickle as do steels but light scale stains can be removed by a bath of 25% nitric and 2% hydrofluoric acid, concentrated nitric-hydrofluoric acid, phosphoric acid, or various bright dips. Deoxidizing of magnesium is accomplished with baths of chromic or 30% hydrofluoric acid at room temperature. Hot sulfuric acid or sulfuric-nitric baths are used on zinc. Titanium is treated with 50% nitric acid and 5% hydrofluoric acid (3).

As mentioned previously, excessive loss of metal may occur in pickling operations. Several inhibitors have been utilized to reduce this attack on the part, including coal tar still residues, guinaldine, substituted pyridines, dibutyl thiorea, ditoyl thiorea, hexamethylenetetramine, and salts of arsenic and molybdenum.

Acid bright dips are usually mixtures of nitric and sulfuric acids employed to provide a mirrorlike surface on cadmium, magnesium, copper, copper alloys, silver, and, in some cases, stainless steel. Nitrogen oxide gases are commonly emitted from these electrolytic baths.

Table 2.2-2 *Contaminants Released by Acid Dip Operations*

Type	Component of Bath That May Be Released to Atmosphere	Physical and Chemical Nature of Major Atmospheric Contaminant
Aluminum bright dip	Phosphoric, nitric acids	Nitrogen oxide gases
Aluminum bright dip	Nitric, sulfuric acids	Nitrogen oxide gases, acid mist
Cadmium bright dip	None	None
Copper bright dip	Nitric, sulfuric acids	Nitrogen oxide gases, acid mist
Copper semibright dip	Sulfuric acid	Acid mist
Copper alloys bright dip	Nitric, sulfuric acids	Nitrogen oxide gases, acid mist
Copper matte dip	Nitric, sulfuric acids	Nitrogen oxide gases, acid mist
Magnesium dip	Chromic acid	Acid mist, steam
Magnesium dip	Nitric, sulfuric acids	Nitrogen oxide gases, acid mist
Monel dip	Nitric, sulfuric acids	Nitrogen oxide gases, acid mist
Nickel and nickel alloys dip	Nitric, sulfuric acids	Nitrogen oxide gases, acid mist
Silver dip	Nitric acid	Nitrogen oxide gases
Silver dip	Sulfuric acid	Sulfuric acid mist
Zinc and zinc alloys dip	Chromic, hydrochloric acids	Hydrogen chloride gas (if HCl attacks Zn)

Source.   Reference 4.

Various acid mists and gases are released from pickling operations (Table 2.2-1) and bright dips (Table 2.2-2) depending on bath temperature, surface area of work, current density (if bath is electrolytic), and whether the bath contains inhibitors that produce a foam blanket on the bath or that lower the surface tension of the bath and thereby reduce misting (4). The ventilation requirements can be calculated for individual processes using the method described in Section 2.4 (4, 5). Due to carryover of pickling acids in the rinse, the rinse tanks may also require local exhaust ventilation. Bright-dip baths require more efficient hooding and exhaust than do most other acid treatment operations.

The minimum safe practices for employees who conduct pickling and bright-dip operations have been proposed by Spring (3).

1   Wash hands and faces before eating, smoking, or leaving plant. Eating and smoking should not be permitted at the work location.

2 Only authorized employees should be permitted to make additions of chemicals to baths.

3 Face shields, chemical handlers' goggles, rubber gloves, rubber aprons, and rubber platers' boots should be worn when adding chemicals to baths, and when cleaning or repairing tanks.

4 Chemicals contacting the body should be washed off immediately and medical assistance obtained.

5 Supervisor should be notified of any change in procedures or unusual occurrences.

2.2.2 Alkaline Treatment

Alkaline immersion cleaning

Acid and alkali cleaning techniques are complementary in terms of the cleaning tasks that can be accomplished. The alkaline immersion systems are used in soak, spray, and electrolytic cleaning and are superior for removal of oil, gases, buffing compounds, certain soils, and paint. A range of alkaline cleansers including sodium hydroxide, potassium hydroxide, sodium carbonate, sodium meta or orthosilicate, trisodium phosphate, borax, and tetrasodium pyrophosphate are used for both soak and electrolytic alkaline cleaning solutions (1, 2).

Table 2.2-3 *Composition of an Alkaline Soak Cleaner for Steel*

Component	Purpose
Caustic soda	Neutralize acidic soil
Sodium silicate	Disperse solids, assist detergency
Trisodium phosphate	Peptize the soil
Tetrasodium pyrophosphate	Detergency action, mineral oil, soils
Sodium tripyrophosphate	Reduce hardness
Surfactants	Reduce surface tension
Soda ash	Alkaline reserve and electrolyte

Source: Reference 3.

The composition of the alkaline bath may be complex, as shown in Table 2.2-3, with a number of additives to handle specific tasks (3). In nonelectrolytic cleaning of rust from steels, the bath may contain 50–80% caustic soda in addition to chelating and sequestering agents. The parts are immersed for 10–15 min and rinsed with a spray; this cycle is repeated until the parts are derusted. Although excellent for rust, this technique will not remove scale.

Electrolytic alkaline cleaning is an aggressive cleaning method and usually follows some primary cleaning. As in electroplating (Section 2.4), the bath is an electrolytic cell powered with direct current with the workpiece conventionally the cathode and an inert electrode as the anode. The water dissociates; oxygen is released at the anode and hydrogen at the cathode. The initial soak is used to loosen the soil; when the system is powered the hydrogen gas generated at the workpiece causes agitation of the surface soils with excellent soil removal.

Conventional steel cleaning is conducted at 550–1700 A/m^2 for a period of $\frac{1}{2}$ to 3 min. Zinc is cleaned at a lower current density of 100–330 A/m^2 for $\frac{1}{3}$ to 1 min. Periodic reversing of the current may be used to enhance cleaning.

As in the case of electroplating baths, the gases generated at the electrodes may result in the release of caustic mist and steam at the surface of the bath. Mist generation is greatest under cathodic cleaning (workpiece is the cathode) and varies with bath concentration, temperature, and current density.

Surfactants and additives that provide a foam blanket on the bath are important to the proper operation of the bath. Ideally, the foam blanket should be 5–8 cm thick to trap the released gas bubbles. The additives must be adjusted to the type of cleaning since soil properties may either enhance or suppress foam generation. If the foam blanket is too thin, the gas may escape causing a significant alkaline mist to become airborne; if too thick, the blanket may trap hydrogen and oxygen with resulting explosions ignited by sparking electrodes (3).

Aluminum is routinely cleaned and etched with strong alkaline solutions in a cathodic cleaning configuration. Due to the surface activity of the solution, significant quantities of the aluminum surface are removed. Surface removal can be controlled to lightly etch the parts, remove surface imperfections, and provide a matte finish. Alkaline baths can be used also to chemically mill aluminum parts to size with removal of large quantities of metal as discussed in Section 2.10.2.

Bath additives in alkaline dip baths for aluminum include triethanolamine and salts of strontium, barium, and calcium to reduce the etch rate and stannous or cobalt salts to increase the etch rate. The surface appearance of the etched part can be modified by the addition of sodium nitrate to the bath.

Salt baths

A bath of molten caustic at 370–540°C (700–1000°F) is used for initial cleaning and descaling of cast iron, alloy, copper, aluminum, and nickel with subsequent quenching and acid pickling (6, 7). The advantages claimed for this type of cleaning are that frequently it does not require precleaning and it provides a good bond for finish. Metal oxides of chromium, nickel, and iron are removed without attacking the base metal; oils are burned off, and graphite, carbon, and sand are removed. The oxides and other debris collect as sludge on the bottom of the bath. The other contaminants combine with the molten caustic, float on the surface of the bath, or are volatized as vapor or fume.

Molten sodium hydroxide is used at 430–540°C (800–1000°F) for general purpose descaling and removal of sand on castings. A modification of this bath, the Virgo process, uses a few percent of sodium nitrate or a chlorate salt to enhance the performance of the bath. Electrolytic processes with sodium hydroxide utilize a two-tank system; the first tank is operated anodically and the part is then removed to a second tank where the workpiece is the cathode.

A reducing process utilizes sodium hydride in the bath at 370°C (700°F) to reduce oxides to their metallic state. The bath utilizes fused liquid anhydrous sodium hydroxide with up to 2% sodium hydride, which is generated in accessory equipment by reacting metallic sodium with hydrogen.

All molten baths require subsequent quenching, pickling, and rinsing. The quenching operation dislodges the scale through steam generation and thermal shock.

These baths require well-defined operating procedures due to the hazard from molten caustic. Local exhaust is necessary and the tank must be equipped with a complete enclosure to protect the operator from violent splashing as the part is immersed in the bath. Quenching tanks and pickling tanks must also be provided with local exhaust ventilation. Ventilation standards have been proposed for these operations (4, 5).

REFERENCES

*1 A. K. Graham, Ed., *Electroplating Engineering Handbook,* 3rd ed., Reinhold, New York, 1971.

 2 James A. Murphy, Ed., *Surface Preparation and Finishes for Metals,* McGraw-Hill, New York, 1971.

*3 S. Spring, *Industrial Cleaning,* Prism Press, Melbourne, 1974.

 4 Committee on Industrial Ventilation, American Conference of Governmental Industrial Hygienists, *Industrial Ventilation: A Manual of Recommended Practice,* 16th ed., ACGIH, Lansing, MI, 1980.

 5 American National Standards Institute, "Standard Practices for Ventilation and Operation of Open-surface Tanks," ANSI Z9.1-77, New York, 1977.

 6 R. H. Shoemaker, *Met. Finish.,* **66,** 60 (1968).

 7 T. Lyman, Ed., *Metals Handbook,* 8th ed., Vol. 2, American Society for Metals, Metals Park, OH, 1964.

2.3 DEGREASING

The removal of surface grime, oil, and grease from metal is commonly done by cold degreasing, soak cleaning, or vapor phase degreasing. The soak cleaners can be straight solvents, emulsion cleaners, or solvent-emulsion cleaners. The significant problems associated with these three degreasing processes are described in the following sections.

2.3.1 Cold Degreasing

Cold degreasing is practiced routinely in small shops and garages for the cleaning of metal parts. The solvents used may vary from high flash point petroleum distillates to mixtures that include aliphatic and aromatic hydrocarbons, chlorinated hydrocarbons, ketones, cellosolves, creosote, and cresylic acid. No generalities regarding control can be made except that no readily volatile or fast-drying solvent should be routinely used in large open containers without effective mechanical exhaust ventilation. Skin contact with these materials should be avoided, and glasses and a face shield should be used to protect the eyes and face from accidental splashing. Covered soak tanks with an adjacent work table with an exhaust hood offer satisfactory vapor control.

When solvents are kept in a safety can or other suitable covered container and applied in small amounts by brushing and wiping, the inhalation hazard far exceeds any fire hazard. Spraying with high flash point petroleum distillates such as Stoddard Solvent, mineral spirits, or kerosene is a widely used method of cleaning oils and grease from metals. Solvents with a flash point below 38°C (100°F) should not be routinely used for this purpose. If this technique is used, the operation should be provided with suitable local exhaust ventilation. The hood may be a conventional spray-booth type and may or may not be fitted with a fire door and automatic extinguishers. The fire hazard attendant to spraying a high flash petroleum solvent is comparable to spraying many lacquers and paints.

2.3.2 Emulsion Cleaners

Emulsion cleaners containing petroleum and coal-tar solvents are commonly used in power washers and soak tanks. When used in soak tanks at room temperature, ventilation is usually not required. When the cleaner is sprayed or used hot, the operation should be confined and provided with local exhaust ventilation. Emulsion cleaners containing cresylic acid, phenols, or halogenated hydrocarbons should also be provided with ventilation and skin contact should be carefully avoided.

2.3.3 Vapor Phase Degreasing

A vapor phase degreaser is a tank containing a quantity of solvent that is heated to its boiling point. The solvent vapor rises and fills the tank to an elevation determined by the location of a condenser. The vapor condenses and returns to the liquid sump. The tank has a freeboard that extends above the condenser to minimize air currents inside the tank (1, 2).

Types of vapor phase degreasers and solvents

The simplest form of vapor phase degreaser, shown in Figure 2.3-1, utilizes only the vapor for cleaning. As the parts are lowered into the hot vapor, the vapor condenses on the cold part and dissolves the surface oils and greases. This oily condensate drops back into the liquid solvent at the base of the tank. The solvent is evaporated continuously to form the vapor blanket. Since the oils are not vaporized, they remain to form a sludge in the bottom of the tank. The scrubbing action of the condensing vapor continues until the temperature of the part reaches the temperature of the vapor whereupon condensation stops, the part appears dry, and it is removed from the degreaser. The time required to reach this point depends on the particular solvent, the temperature of the vapor, the weight of the part, and its specific heat. The vapor phase degreaser does an excellent job of drying parts after aqueous cleaning and before plating and is frequently used for this purpose in the jewelry industry.

The straight vapor-cycle degreaser is not effective on small, light work since the part reaches the temperature of the vapor before the condensing action has cleaned the part. For such an application the vapor-spray cycle degreaser shown in Figure 2.3-2 is frequently used. The part to be cleansed is first placed in the vapor zone as in the straight vapor cycle degreaser. A portion of the vapor is condensed by a cooling coil and fills a liquid solvent reservoir. This warm liquid solvent is pumped to a nozzle that can be used to direct the solvent on the part, washing off surface oils and cooling the part, thereby permitting final cleaning by vapor condensation.

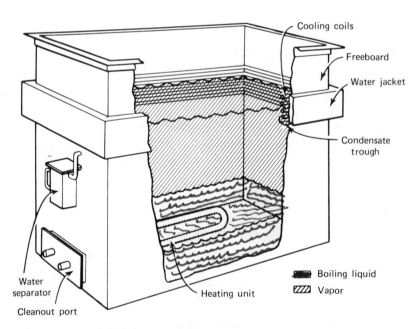

Figure 2.3-1 Straight vapor cycle degreaser. (From *Surface Preparation and Finishes For Metals* by J. A. Murphy. Copyright 1971. Used with permission of McGraw-Hill Book Co.)

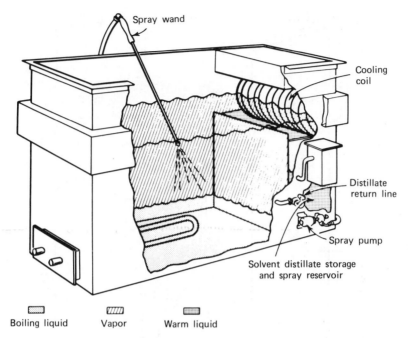

Spray wand

Cooling coil

Distillate return line

Spray pump

Solvent distillate storage and spray reservoir

Boiling liquid   Vapor   Warm liquid

Figure 2.3-2 Vapor-spray cycle degreaser. (From *Surface Preparation and Finishes For Metals* by J. A. Murphy. Copyright 1971. Used with permission of McGraw-Hill Book Co.)

A third degreaser design, shown in Figure 2.3-3, has a compartment with warm solvent and a second compartment with a vapor zone. This degreaser is used for heavily soiled parts or to clean a basket of small parts that nest together. The other major vapor phase degreaser design, shown in Figure 2.3-4, has both boiling and warm liquid compartments. The boiling liquid compartment is installed to maintain the vapor phase when parts of high heat capacity are processed through the unit. Other specialty degreasers encountered in industry include enclosed conveyorized units for continuous production cleaning.

The popular solvents used with vapor phase degreasers include trichloroethylene, perchloroethylene, methyl chloroform, methylene chloride, and trichlorotrifluoroethane (Table 2.3-1) (3, 4, 5, 6). The degreaser must be designed for and used with a specific solvent. Most commercial degreasing solvents sold under trade names contain a small amount of stabilizer. The purpose of the stabilizer is to neutralize any free acid that might result from oxidation of the degreasing liquid in the presence of air, hydrolysis in the presence of water, or pyrolysis under the influence of high temperatures.

At one time the use of ultrasonic degreasers was restricted to critical degreasing operations. Increasingly they are being used for more conventional cleaning jobs due to cleaning efficiency and speed. An ultrasonic degreaser,

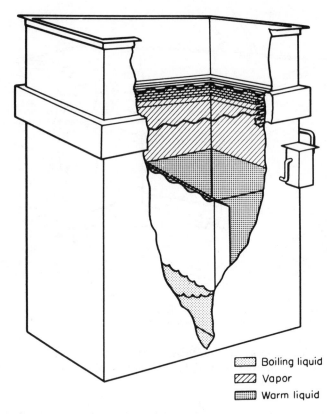

Boiling liquid

Vapor

Warm liquid

Figure 2.3-3 Liquid-vapor cycle degreaser. (From *Surface Preparation and Finishes For Metals* by J. A. Murphy. Copyright 1971. Used with permission of McGraw-Hill Book Co.)

Table 2.3-1 *Properties of Vapor Degreasing Solvents*

	Trichloroethylene	Perchloro-ethylene	Methylene Chloride	Trichloro trifluoro ethane[a]	Methyl Chloroform (1,1,1 Tri chloroethane)
Boiling point					
°C	87	121	40	48	74
°F	188	250	104	118	165
Flammability	Nonflammable under vapor degreasing conditions				
Latent heat of vaporization					
(b.p.), Btu/lb	103	90	142	63	105
Specific gravity					
Vapor (Air = 1.00)	4.53	5.72	2.93	6.75	4.60
Liquid (Water = 1.00)	1.464	1.623	1.326	1.514	1.327

[a]Binary azeotropes are also available with ethyl alcohol, isopropyl alcohol, acetone, and methylene chloride.

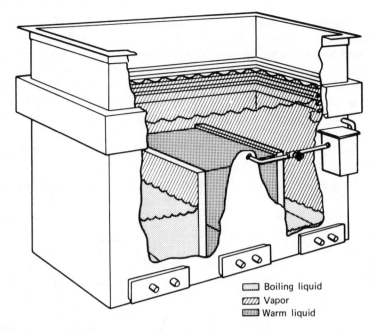

Boiling liquid
Vapor
Warm liquid

Figure 2.3-4 Liquid-liquid-vapor cycle degreaser. (From *Surface Preparation and Finishes For Metals* by J. A. Murphy. Copyright 1971. Used with permission of McGraw-Hill Book Co.)

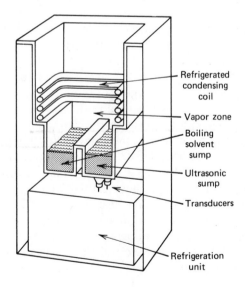

Refrigerated condensing coil

Vapor zone

Boiling solvent sump

Ultrasonic sump

Transducers

Refrigeration unit

Figure 2.3-5 Ultrasonic degreaser. (Reproduced courtesy of Branson Cleaning Equipment Co.)

shown in Figure 2.3-5, has a transducer mounted at the base of the tank which operates in the range of 20 to 40 kHz (7). The transducer alternately compresses and expands the solvent forming small bubbles that cavitate or collapse at the surface of the workpiece. The cavitation phenomenon disrupts the adhering soils and cleans the part. Ultrasonic degreasers use chlorinated solvents at 32–49°C (90–120°F) and aqueous solutions at 43–71°C (110–160°F). The cleaning solutions recommended by one manufacturer of ultrasonic equipment are shown in Table 2.3-2 (7). These degreasers commonly employ refrigerated or water-chilled coils for control of solvent vapors and the manufacturers claim that local exhaust ventilation is not needed.

The heat input to the conventional vapor phase degreaser is commonly provided by electricity or steam, although older units utilize indirect combustion heaters fueled by gas, kerosene, or fuel oil. The recent emphasis on energy conservation has had an impact on the choice of solvents for vapor phase degreasers. To minimize energy, one must consider the boiling temperature, specific heat, and latent heat of vaporization of the candidate solvents. The chlorinated degreasing solvents (trichloroethylene, methylene chloride, 1,1,1 trichloroethane, and perchloroethylene) are still the major degreasing solvents in use, however, fluorocarbons are making inroads due to the low energy requirements necessary to boil the solvent and provide the vapor zone. Du Pont claims low solvent consumption for the Freon ® degreasing solvents (4). One claim is that large parts introduced into the vapors of the high boiling point chlorinated solvents cause rapid vapor condensation on the part with the result that the solvent cannot be boiled rapidly enough to replace the vapor. Air is then pulled into the degreaser only to be displaced into the workplace as solvent-laden air when the vapor blanket is replaced. A second reason claimed for reduced solvent losses with Freons is that with high boiling chlorinated solvents it takes the part a long time to attain thermal equilibrium with the vapor, prompting early removal of the part by an impatient operator with resulting "carry out" of solvent with the part. It is stated that emissions from a degreaser operated with Freons are so low that local exhaust ventilation is not required.

The loss of degreaser solvent to the workplace obviously depends on a number of operating conditions, including the cleaning cycle, work load, and the type of material being cleaned. The loss of solvent from an idling open-top degreaser (60 by 140cm) located in an area without significant drafts expressed as pounds per square foot per hour is 0.14 for 1,1,1 trichlorethane, 0.20 for trichloroethylene, 0.26 for methylene chloride, and 0.29 for perchloroethylene (8). Airborne concentrations at the operator's breathing zone measured by a direct reading instrument during active operation of degreasers vary widely, as shown in Table 2.3-3 (8). Time weighted average concentrations for a full 8 hr shift developed by a state agency based on personal air sampling are shown in Table 2.3-4.

Table 2.3-2 *Cleaning Solutions for Ultrasonic Degreasers*

Solvents	Commercial Equivalent
Trichlorotrifluorethane	Freon TF
Trichlorotrifluoroethane and acetone, azeotrope of	Freon TA
Trichlorotrifluoroethane and ethanol, azeotrope of	Freon TE
Trichlorotrifluoroethane and methylene chloride, azeotrope of	Freon TMC
Trichlorotrifluoroethane–water emulsion	Freon TWD-602
Methylene chloride	Methylene chloride
1,1,1 Trichloroethane	Chlorothene VG
Perchloroethylene	Perchloroethylene
Special blends	—
Aqueous	
Acidic	Oakite 31
Acidic, inhibited	Oakite 131
Acidic, plus solvent	Oakite 33
Chelated Detergents	
Neutral (pH 5–8)	Oakite Renovator
Mild alkaline (pH 8–10)	Oakite FM 403
Strong alkaline (pH 11–14)	Oakite Rustripper
Nonchelated	
Neutral detergent (pH 5–8)	Joy
Mild alkaline detergent (pH 8–10)	Oakite 202 or Liqui-Det
Strong alkaline detergent (pH 11–14)	Oakite Versadet or Oakite 162
Others	
Inhibited alkaline cleaner	Oakite Aluminum Cleaner 166
Soap-free alkaline	Oakite 12
Special stripper	Oakite Stripper 156

Source: Reference 7. Reproduced Courtesy of Branson Cleaning Equipment Co.

Table 2.3-3 *Typical Open-Top Vapor Degreaser Air Measurements[a]*

Location of Measurement	Trichloro-ethylene (ppm)	1,1,1-Tri-chloroethane (ppm)	Perchloro-ethylene (ppm)	Methylene Chloride (ppm)
Breathing zone of operator when lowering parts into degreaser	139–259	91–199	74–174	166–276
Breathing zone of operator with parts in vapor zone	201–439	138–257	100–303	63–372
Breathing zone of operator during removal of parts from degreaser	236–466	168–364	190–526	171–786

Source. Reference 8. Reproduced courtesy of The Dow Chemical Co.

[a]Data for trichloroethylene, 1,1,1-trichloroethane, and perchloroethylene are based on surveys of some 1000 degreasers. Data for methylene chloride are based on more limited surveys.

Table 2.3-4 *Full Shift TWA Concentrations During Vapor Degreasing*

Solvent	No. of Measurements	Concentration (ppm) Range	Average
Trichloroethylene	14	5–311	62
Perchloroethylene	6	5–174	40

Control (1, 9, 10)

Vapor phase degreasers should have adequate condensers in the form of water jackets, pipe coils, or both, extending around the tank. The condenser prevents the escape of the concentrated vapors into the room. The vertical distance between the lowest point at which vapors can escape from the degreaser machine and the highest normal vapor level is called the freeboard. The freeboard should be at least 40 cm and not less than half the width of the machine. The portion of the condenser above the vapor line should be maintained above room temperature and below 43°C (110°F). The effluent water should be regulated to this same range, and a temperature indicator or control is desirable. A proprietary refrigerated device was introduced in the 1970s and has seen limited application on vapor phase degreasers (11). On this device finned refrigeration lines operating at −29°C (−20°F) are located on the inside of the degreaser directly above the vapor line. The manufacturer of the refrigerated condenser states that the unit provides a cold air blanket across the top of the working vapor zone reducing the loss of solvent vapor and eliminating the need for local exhaust ventilation. Data supporting this claim provided by the manufacturer are shown in Table 2.3-5.

All machines should have thermostatic controls located a few inches above the normal vapor level to shut off the source of heat if the vapor rises above the condensing surface and a boiling liquid thermostat to prevent overheating. The recommended temperatures for these thermostats are shown in Table 2.3-6.

There is a difference of opinion on the need for local exhaust ventilation on vapor phase degreasers. Ventilation control may be required depending on the location, maintenance, and operating practices; however, unless it is properly designed it may increase operator exposure. Local exhaust will certainly increase solvent loss and the installation may exceed air pollution emission limits requiring solvent recovery before discharging to outdoors. To ensure effective use of local exhaust, the units should be installed away from drafts from open windows, spray booths, space heaters, supply air grilles, and fans (12). The parts loading and unloading station is equally important. When baskets of small parts are degreased, it is not possible to completely eliminate drag-out and the unloading station may require ventilation control.

When degreasers are installed in pits, mechanical exhaust ventilation should be provided at the lowest part of the pit. Open flames, electric heating elements, and welding operations should be divorced from the degreaser locations, since the solvent will be degraded by both direct flame and ultraviolet radiation, thereby producing toxic air contaminants.

Table 2.3-5 *Performance of Cold Trap*

Operating Condition		Average Concentration at Operator's Working Position (ppm) (solvent not named)
Exhaust	Cold Trap	
Off	Off	332
On	Off	13
Off	On	3

Source.   Reference 11.

Table 2.3-6 *Thermostat Settings*

Solvent	Safety Vapor Control Thermostat Setting		Safety Thermostat for Boiling Liquid	
	(°C)	(°F)	(°C)	(°F)
Methylene chloride	35	95	52	125
Perchloroethylene	104	220	145	295
1,1,1 Trichloroethane	54	130	88	190
Trichloroethylene	71	160	116	240
Trichlorotrifluoroethane	40	105	53	130

Source.   Reference 3.

In Section 2.15, reference is made to the decomposition of chlorinated solvents under thermal and ultraviolet stress with the formation of chlorine, hydrogen chloride, and phosgene. Since degreasers using such solvents are frequently located near welding operations, this problem warrants attention. In a laboratory study of the decomposition potential of methyl chloride, methylene chloride, carbon tetrachloride, ethylene dichloride, 1,1,1 trichloroethane, o-dichlorobenzene, trichloroethylene, and perchloroethylene, only the latter two solvents decomposed in the welding enviroment to form dangerous levels of phosgene, chlorine, and hydrogen chloride (Figure 2.3-6). Certain chlorinated materials will also degrade if introduced to direct fired combustion units commonly used in industry. If a highly corroded heater is noted in the degreaser area, it may indicate that toxic and corrosive air contaminants are being generated.

Installation instructions and operating precautions for the use of conventional vapor phase degreasers have been proposed by various authorities. The following instructions should be observed for all degreasers:

1   If the unit is equipped with a water condenser, the water should be turned on before the heat is applied to the solvent.

2   Water temperature should be maintained between 27°C (81°F) and 43°C (110°F).

3   Work should not be placed in and removed from the vapor faster than 0.055 m/s (11 fpm). If a hoist is not available, a support should be positioned

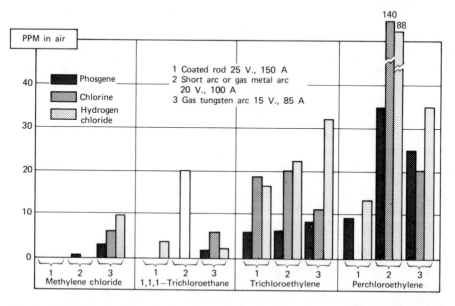

Figure 2.3-6 Phosgene, chlorine, and hydrogen chloride releases near vapor phase degreasers. (Reproduced courtesy of the Dow Chemical Co.)

to hold the work in the vapor. This minimizes the time the operator must spend in the high exposure zone.

4 The part must be kept in the vapor until it reaches vapor temperature and is dry.

5 Parts should be loaded to minimize pullout. For example, cup-shaped parts should be inverted.

6 Overloading should be avoided because it will cause displacement of vapor into the workroom.

7 The work should be sprayed below the vapor level.

8 Proper heat input must be available to ensure vapor recovery when large loads are placed in the degreaser.

9 A thermostat should be installed in the boiling solvent to prevent overheating of the solvent.

10 A thermostat vapor level control must be installed on the degreaser and set for the particular solvent in use.

11 The degreaser tank should be covered when not in use.

12 Hot solvent should not be removed from the degreaser, nor should garments be cleaned in the degreaser.

13 An emergency eye wash station should be located near the degreaser for prompt irrigation of the eye in case of an accidental splash.

Vapor phase degreasers should be cleaned periodically to prevent the accumulation of sludge and metal chips. The solvent should be distilled off until the heating surface or element is nearly but not quite exposed or until the solvent vapors fail to rise to the collecting trough. After cooling, the oil and solvent should be drained off and the sludge removed. It is important that the solvent be cooled before draining. Removing hot solvent causes serious air contamination and frequently requires the evacuation of plant personnel from the building. A fire hazard may exist during the cleaning of machines heated by gas or electricity because the flash point of the residual oil may be reached and because trichloroethylene itself is flammable at elevated termperatures. After sludge and solvent removal, the degreaser must be ventilated mechanically before any maintenance work involving flames or welding is undertaken. A person should not be permitted to enter a degreaser or place his or her head in one until all controls for entry into a confined space have been put in place. Anyone entering a degreaser should wear a respirator suitable for conditions immediately hazardous to life, as well as a lifeline held by an attendant. In such circumstances, anesthetic concentrations of vapor may be encountered and oxygen concentration may be insufficient. Such an atmosphere may cause unconsciousness with little or no warning. A number of deaths have occurred because of failure to observe the foregoing precautions (13).

As mentioned previously, trichloroethylene is flammable at the elevated temperatures present in degreasers. When heated above 43°C (110°F), the solvent has a narrow flammable range around 20% by volume. This range increases with temperature, and above 63°C (135°F) the flammable range is from 15 to 40% by volume. The ignition temperature is 410°C (770°F). These conditions do not ordinarily occur in plant atmospheres but may occur within a degreaser. Trichloroethylene vapors will not explode violently under any circumstances, but they may burn slowly to form dense smoke and gases such as chlorine, hydrogen chloride, and phosgene. Although perchloroethylene vapor will not ignite or burn, oils or greases accumulated in the degreaser will; therefore, sources of ignition, especially overheating with gas or electric heaters, should be avoided during distillation for sludge removal. Also, welding on or in a degreaser when it contains solvent should be avoided.

The substitution of one degreaser solvent for another as a control technique must be done with caution. Such a decision should not be based solely on the relative TLVs but must consider the total toxicology of the solvents, their photochemical properties, and the physical properties of the solvents including vapor pressure.

Although it is frequently stated that vapor phase degreasers do not require local exhaust ventilation, my experience has been that the location, maintenance, and operation of most degreasers do result in exposures that warrant such control. Equally important is the parts loading and unloading station.

Medical control is especially important for degreasing operators and should include both preplacement and periodic physical examinations. Specific medical screening is advisable for persons with cardiovascular disease.

Cardiac arrhythmias
other hazards - Asphyxia, drying + defatting,
release of break down by products

REFERENCES

*1 American Society for Testing and Materials, *Handbook of Vapor Degreasing*, STP 310A, ASTM, Philadelphia, PA, 1976.

2   J. A. Murphy, Ed., *Surface Preparation and Finishes for Metals*, McGraw-Hill, New York, 1971.

3   "Modern Vapor Degreasing and Dow Chlorinated Solvents," Dow Bulletin, Form No. 100-5185-77, Dow Chemical Co., Midland, MI, 1977.

4   "Metals Can Be Cleaned with Less Energy," Bulletin Form E-14217, 6/77 IOM, DuPont, Wilmington, DE, 1977.

5   "Freon Solvent Data," Bulletin No. FST-3, DuPont, Wilmington, DE, 1978.

6   T. J. Kearney, *Metal Finishing Guidebook Directory,* Metals and Plastics Publications, Inc., Westwood, NJ, 1970.

7   Technical Bulletins, Branson Cleaning Equipment Co., Shelton, CT.

8   "How To Select a Vapor Degreasing Solvent," Bulletin Form 100-5321-78, Dow Chemical Co., Midland, MI, 1978.

9   "Vapor Degreasing," Industrial Hygiene Series No. 29, American Mutual Insurance Alliance, Chicago, IL, 1973.

10 "General Instruction Manual," Bulletin 1S 6608.2, Detrex Chemical Industries, Detroit, MI, 1966.

11 "The Cold Trap, Freeboard Chiller Device," AutoSonics, P.O. Box 300, Conshohocken, PA, 1972.

12 Committee on Industrial Ventilation, American Conference of Governmental Industrial Hygienists, *Industrial Ventilation: A Manual of Recommended Practice*, 16th ed., ACGIH, Lansing, MI, 1980.

13 M.W. First, *J. Am. Soc. Saf. Eng.* **14,** 11 (1969).

2.4 ELECTROPLATING

Metal, plastic, and rubber parts are plated to prevent rusting and corrosion, for appearance, to reduce electrical contact resistance, to provide electrical insulation, as a base for soldering operations, and to improve wearability. The common plating metals include cadmium, chromium, copper, gold, nickel, silver, and their alloys. Prior to electroplating the parts must be cleaned and the surfaces treated as described in Sections 2.2 and 2.3.

There is no published estimate of the number of electroplaters at work in the United States or the distribution between "captive" electroplating facilities and independent shops. It is believed that the majority of platers are employed by independent shops and that the majority of these shops employ less than 10 persons. The potential health hazards in electroplating operations are numerous; however, a comprehensive epidemiologic study of this industry in the United States has not been completed so that the impact of the health hazards on the workers is not known. A recent survey of cancer mortality in U.S. counties with more than 0.1% of the population engaged in the electroplating and coating industry has shown excess mortality rate for cancers of the esophagus and the larynx (1).

2.4.1 Electroplating Techniques

The basic electroplating system is shown in Figure 2.4-1. The plating tank contains an electrolyte consisting of a metal salt of the metal to be applied

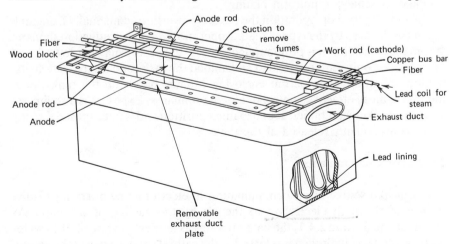

Figure 2.4-1 Electroplating tank.

dissolved in water. Two electrodes powered by a low voltage dc power supply are immersed in the electrolyte. The cathode is the workpiece to be plated and the anode is either an inert electrode or a bar of the metal to be deposited. When power is applied, the metal ions deposit out of the bath on the cathode or workpiece. Water is dissociated releasing hydrogen at the cathode and oxygen at the anode. The anode may be designed to replenish the metallic ion concentration in the bath. Current density expressed in amperes per unit area of workpiece surface varies depending on the operation. In addition to the salt containing the metallic ion, the plating bath may contain salts to adjust the electrical conductivity of the bath, additives that determine the type of plating deposit, and a buffer for pH control of the bath.

Anodizing, a common surface treatment for decoration, corrosion resistance, and electrical insulation on such metals as magnesium, aluminum, and titanium, operates in a different fashion. The workpiece is the anode, and the cathode is a lead bar. The oxygen formed at the workpiece causes a controlled surface oxidation. The process is conducted in a chromic acid bath with high current density and since its efficiency is quite low, the amount of misting is high.

Several touch-up electroplating procedures, with spot chrome the most common, are available now. In this process the conventional chrome plating procedure is conducted with small quantities of chemicals using an electrode-pad system. The repair processes are hazardous in that there is a good chance for skin contact in preparing and handling the chemically saturated pads, and the acids and cyanide chemicals may be mixed inadvertently.

In the conventional plating operations, individual parts or racks of small parts are hung manually from the cathode bar. If many small pieces are to be plated, the parts are placed in a perforated plastic barrel in electrical contact with the cathode bar and the barrel is immersed in the bath. The parts are tumbled to achieve a uniform plating.

In a small job-shop operation the parts are transferred manually from tank to tank as dictated by the type of plating operation. In high production shops an automatic transfer unit is programmed to cycle the parts from tank to tank and the worker is only required to load and unload the racks or baskets. Automatic plating operations may permit enclosing exhaust hoods on the tanks, and, therefore, more effective control of air contaminants. Exposure is also limited since the worker is stationed at one loading position and is not exposed directly to air contaminants released at the tanks.

2.4.2 Air Contaminants

The principal source of air contamination in electroplating operations is the release of the bath electrolyte to the air by the gassing of the bath. As mentioned in Section 2.4.1, the bath operates as an electrolytic cell; thus water is dissociated and hydrogen is released at the cathode and oxygen at the anode.

The gases released at the electrodes rise to the surface of the bath and break, generating a fine mist which becomes airborne. The presence of the mist can be detected by placing a clean piece of paper parallel to the bath approximately 2 cm above the surface. The mist generation rate depends on the bath efficiency. As shown in Table 2.4-1, in copper plating the efficiency of the sulfate plating bath is nearly 100%; that is, essentially all the energy goes into the plating operation and little goes into the dissociation of water (2). Nickel plating baths operate at 95% efficiency so that only 5% of the energy is directed to dissociation of water and gassing is minimal. However, chromium plating operations are quite inefficient and up to 90% of the total energy may be devoted to dissociation of the bath with resulting severe gassing and potential exposure of the operator to chromic acid mist. Although the contaminant generation rate of the bath is governed principally by the efficiency of the bath, it also varies with the metallic ion concentration, the current density, the nature of bath additives, and bath temperature. Air or mechanical agitation of the bath, used to improve plating quality, may also release the bath as a mist.

Table 2.4-1 *Electroplating Bath Operating Conditions*

Deposited Metal	Application	Bath Composition	Temperature (°C)	Cathode Current Density (A/m^2)	Cathode Current Efficiency (%)
Cadmium	Decorative Protective	Cyanide	30	5–50	90–95
Chromium	Hard plate	Dilute CrO$_3$	45–55	50–200	10–20
	Decorative	Conc. CrO$_3$	40–50	80–120	10–20
Copper	Strike	Cyanide	40–60	10–40	30–50
	Heavy plate	Sulfate	20–50	20–100	95–100
Gold	Electrical Decorative	Cyanide	60–70	1–10	95
Nickel	Protective	Watts	55	10–60	95
		Chloride	55	50–100	95
		Fluoborate	55	50–100	95
Silver	Strike Cyanide	Cyanide strike	20	30	99
	heavy	Cyanide heavy	20–50	10–150	99
Tin	Tin plate	Acid sulfate (1)	25–30	10–60	95
		(2)	40–50	100–400	95
		Alkaline (1)	80	10–60	95
		(2)	85	100–400	95
Zinc	Brightener	Cyanide	30–50	10–50	75–95
	Brightener	Sulfate	25–35	10–30	95

Source. Reference 2. Courtesy of Pergamon Press Ltd.

The health significance of the mist generated by electroplating processes depends, of course, on the contents of the bath. An inventory of the nature of the chemicals in the common electroplating baths, the form in which they are released to the air, and the rate of gassing has evolved over the past decades drawing heavily on the experience of the industrial hygiene programs in the states of New York and Michigan (3, 4). These data, shown in Table 2.4-2, are useful in defining the nature of the contaminant and the air sampling protocol necessary to define the worker exposure. As will be noted later in this discussion, these data are also valuable in defining the ventilation requirements for various plating operations.

Table 2.4-2 *Contaminants Released by Electroplating Operations*

Process	Metal	Component of Bath That May Be Released to Atmosphere	Physical and Chemical Nature of Major Atmospheric Contaminant
Electrodeless plating	Copper	Formaldehyde	Formaldehyde gas
	Nickel	Ammonium hydroxide	Ammonia gas
Alkaline electroplating	Platinum	Ammonium phosphate, ammonia gas	Ammonia gas
	Tin	Sodium stannate	Tin salt mist, steam
	Zinc	None	None
Fluoborate electroplating	Cadmium	Fluoborate salts	Fluoborate mist, steam
	Copper	Copper fluoborate	Fluoborate mist, steam
	Indium	Fluoborate salts	Fluoborate mist, steam
	Lead	Lead fluoborate-fluoboric acid	Fluoborate mist, hydrogen fluoride gas
	Lead-tin alloy	Lead fluoborate-fluoboric acid	Fluoborate mist
	Nickel	Nickel fluoborate	Fluoborate mist
	Tin	Stannous fluoborate, fluoboric acid	Fluoborate mist
	Zinc	Fluoborate salts	Fluoborate mist, steam
Cyanide electroplating	Brass, bronze	Cyanide salts, ammonium hydroxide	Cyanide mist, ammonia gas
	Bright zinc	Cyanide salts, sodium hydroxide	Cyanide, alkaline mists
	Cadmium	None	None
	Copper	None	None
	Copper	Cyanide salts, sodium hydroxide	Cyanide, alkaline mists, steam
	Gold	Cyanide salts	Cyanide mist, steam
	Indium	Cyanide salts, sodium hydroxide	Cyanide, alkaline mists
	Silver	None	None
	Tin-zinc alloy	Cyanide salts, potassium hydroxide	Cyanide, alkaline mists, steam
	White alloy	Cyanide salts, sodium stannate	Cyanide, alkaline mists
	Zinc	Cyanide salts, sodium hydroxide	Cyanide, alkaine mists
Acid electroplating	Chromium	Chromic acid	Chromic acid mist
	Copper	Copper sulfate, sulfuric acid	Sulfuric acid mist
	Gold	Cyanide salts	Cyanide mist
	Indium	None	None
	Indium	Sulfamic acid, sulfamate salts	Sulfamate mist
	Iron	Chloride salts, hydrochloric acid	Hydrochloric acid mist, steam
	Iron	None	None
	Nickel	Ammonium fluoride, hydrofluoric acid	Hydrofluoric acid mist
	Nickel and black nickel	None	None
	Nickel	Nickel sulfate	Nickel sulfate mist
	Nickel	Nickel sulfamate	Sulfamate mist
	Palladium	None	None
	Rhodium	None	None
	Tin	Tin halide	Halide mist
	Tin	None	None
	Zinc	Zinc chloride	Zinc chloride mist
	Zinc	None	None

Source. Reference 1.

2.4.3 Control

Proprietary bath additives are available to reduce the surface tension of the electrolyte and, therefore, to reduce misting. The minimum effective viscosity is 35 dynes/cm but one should operate at a viscosity of 25 dynes/cm (5, 6). Another additive provides a thick foam that traps the mist released from the bath. This agent is best used for tanks that operate continuously. A layer of plastic chips, beads, or balls on the surface of the bath will also trap the mist and permit it to drain back into the bath. Where possible, tanks should be provided with covers to reduce bath loss.

Although these mist suppressants are helpful, they cannot be considered the principal control measure for airborne contaminants from plating tanks. That distinction must rest with local exhaust ventilation (7). The common exhaust hoods on electroplating tanks are end takeoff, lateral slot, or upward plenum slotted hoods.

Until the 1950s it was considered good practice to exhaust all tanks containing nitric acid, chromic acid, hydrofluoric acid, hot cyanide and alkaline solutions, and hot water. The rule of thumb was to provide $0.5 - 0.75$ m³/s per square meter (100–150 cfm per square foot) of tank area. In some cases this was quite adequate; in some cases, it was not. The evolution of the design approach described in Reference 3 has permitted improved control of electroplating air contaminants.

This procedure permits one to determine a minimum control velocity based on the hazard potential of the bath and the rate of contaminant generation derived from Table 2.4-2. The exhaust volume is calculated using this control velocity and the tank measurements and geometry. The reader is referred to Reference 3 for a detailed presentation of the design ventilation procedure; however, the following general observations on plating room exhaust ventilation are helpful (8).

1 Copper plating in an acid solution is normally conducted with low current densities and does not use air agitation. In such cases local exhaust ventilation is not required, but when high current densities or air agitation are used, significant air contamination occurs and the tank must be exhausted.

2 When a cyanide bath is used for strike and bright copper plate, both cyanide and alkali mist are released, and ventilation is required.

3 Neither sulfate nor chloride plating baths for nickel require exhaust ventilation since gassing is negligible.

4 As in the case of acid copper plate, neither zinc nor cadmium plating solutions require exhaust ventilation unless high current densities and bath temperature exist.

In addition to the proper design and installation of good local exhaust ventilation, one must provide adequate makeup air, backflow dampers on combustion devices to prevent carbon monoxide contamination of the workplace, and suitable air cleaning on chromic acid and alkali mist releases.

The difference in exposure to airborne chromium compounds between hard and bright chromium plating has been investigated (9). This study demonstrates that while significant exposures to chromium may exist in hard chromium electroplating, the health hazard from bright chromium electroplating is limited.

As is the case in all industrial processes, effectiveness of exhaust ventilation may be evaluated by air sampling at the workplace and direct ventilation measurements at the bath. Due to the severe corrosion of duct work, periodic checks of the exhaust systems in plating shops are necessary. One should determine the exhaust volumes from each tank by Pitot measurements and compare the observed values with the recommended exhaust rate as calculated using the procedure described previously. Qualitative assessment of the ventilation is possible using smoke tubes or other tracers. It is common practice to measure the hood slot velocity, average the readings, and calculate the exhaust volume from these measurements. This technique has limited value due to the difficulty in defining average velocity in a narrow slot. Measurement of slot velocity is useful in determining the uniformity of the exhaust over the length of the tank, but that is all. The effects of room drafts on capture velocity should be identified. In many cases the use of partitions to minimize the disruptive effects of drafts may greatly improve the installed ventilation.

One would expect that an electrical hazard would exist at the plating tanks. This is not true due to the low voltage dc power supply in use. There is also limited fire and explosion potential in most plating shops. The major chemical safety hazards result from the handling of concentrated acids and alkalies when preparing the baths and the accidental mixing of acids with cyanides or sulfides during plating, bath preparation, and waste disposal with the formation of hydrogen cyanide or hydrogen sulfide.

The educated use of protective equipment by electroplaters is extremely important in preventing contact with the various sensitizers and corrosive materials encountered in the plating shop (6). The minimum protective clothing should include rubber gloves, aprons, boots, and chemical handler's goggles. Aprons should come below the top of the boots. All personnel should have a change of clothing available at the workplace. If solutions are splashed on the work clothing it should be removed, the skin washed, and the worker should change to clean garments. A shower and eye wash station serviced with tempered water should be available at the workplace. The wide range of chemicals handled in an open fashion does present a major dermatitis hazard to the plater, and skin contact must be avoided. Nickel is a skin sensitizer and may cause nickel itch developing into a rash with skin ulcerations (10). One is impressed by the ability of the worker in small shops to be literally immersed in bath fluids without major difficulties. Skin problems do occur, however, if good housekeeping and personal cleanliness are not followed.

A summary of the health hazards encountered in electroplating shops and the available controls is presented in Table 2.4-3 (11).

Table 2.4-3 *Summary of Major Electroplating Health Hazards*

Inhalation
Mist, gases, and vapors
Hydrogen cyanide	Accidental mixing of cyanide solutions and acids
Chromic acid	Released as a mist during chrome plating and anodizing
Hydrogen sulfide	Accidental mixing of sulfide solutions and acids
Nitrogen oxides	Released from pickling baths containing nitric acid

Dust
Released during weighing and transfer of various solid bath materials including cyanides and cadmium salts
Fumes
Generated during on-site repair of lead lined tanks using torch burning techniques

Ingestion
Accidental ingestion of particulate contamination during smoking and eating at workplace.

Skin Contact
Absorption of cyanide compounds through the skin
Defatting by solvents
Primary irritants
Sensitizers

Control Technology
Local exhaust ventilation
Mist reduction
  Reduce surface tension
  Coat surface
  Tank covers
Isolation of stored chemicals

REFERENCES

1 A. Blair and T. Mason, *Arch. Environ. Health,* **35,** 92 (1980).

2 D.R. Blair, *Principles of Metal Surface Treatment and Protection,* Pergamon, Oxford, 1972.

*3 Committee on Industrial Ventilation, American Conference of Governmental Industrial Hygienists, *Industrial Ventilation: A Manual of Recommended Practice,* 16th ed., ACGIH, Lansing, MI, 1980.

4 American National Standards Institute, "Fundamentals Governing the Design and Operation of Local Exhaust Systems," ANSI Z9, New York, 1979.

5 G. Hama, W. Frederick, D. Millage, and H. Brown, *Am. Ind. Hyg. Assoc. J.,* **15,** 211 (1954).

6 H. F. Henning, *Ann. Occup. Hyg.,* **15,** (1972).

7 L. J. Flanigan, S. G. Talbert, D. E. Semeones, and B. C. Kim, "Development and Design Criteria for Exhaust Systems for Open Surface Tanks," NIOSH Research Report Contract No. HSM 099-71-61, Battelle-Columbus Laboratories, Columbus, OH, 1974.

8 "Ventilation for Electroplating Plants," *Mich. Occup. Health,* **12,** 4 (Summer 1967).

9 M. P. Guillemin and M. Berode, *Ann. Occup. Hyg.,* **21,** 105 (1978).

10 A. A. Fisher and A. Shapiro, *J. Am. Med. Assoc.,* **161,** 717 (1956).

*11 A. K. Graham, Ed., *Electroplating Engineering Handbook,* 3rd ed., Reinhold, New York, 1970.

2.5 FORGING

The practice of forging, the plastic deformation of metal to a given size and shape, imparts certain desirable metallurgical properties to the workpiece that cannot be obtained in any other fashion. Industry could not function as it now does without forged parts. A wide range of metals and alloys are processed by cold and hot forging. Since hot forging presents the major occupational health hazards, it will be given attention in this discussion. More than one-half the metal worked is low carbon steel, low alloy metals, and aluminum; the balance of the metals worked include a range of nickel-chromium base alloys, copper, brass and titanium.

2.5.1 Forging Practice

Forging processes are classified by the type of equipment used to form the metal part in an open or closed impression die (1, 2). In drop hammer forging (Figure 2.5-1) the bottom half of the die is positioned on an anvil isolated from the frame of the hammer. The top half of the die is attached to a vertical ram that is raised by an air or steam cylinder and then dropped by gravity or driven downward by air or steam. An operator controls the force of the impact and the frequency of the blows.

The second type of forging utilizes a forging press. In this case, the top die is powered by a mechanical or hydraulic activated press similar in design to a punch press. Presses are used on both open and impression die forging. Upset forging is another technique in which the workpiece is placed on the bottom flat die and its height is reduced by downward movement of the top die.

A series of operations shown in Figure 2.5-2 are conducted in hot forging. Metal stock is cut to size, the part is heated to forging temperature, and the workpiece is forged between heated die blocks. If single cavity dies are used, the part might be cycled between reheat furnaces and a progression of dies mounted on a series of hammers or presses. For small parts with simple geometry, the necessary impressions or cavities may be cut in a single die and the forging can be completed at one hammer or press without reheating.

Between hammer blows, lubricants are applied by swab or spray to the die face and to the workpiece positioned in the bottom die. The die lubricant is

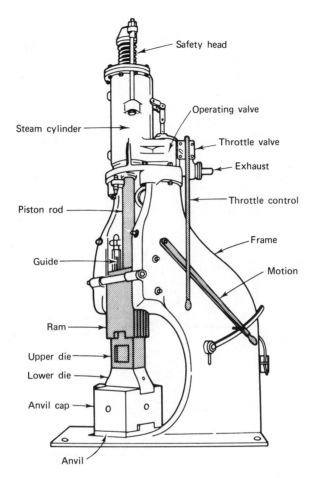

Safety head

Operating valve

Steam cylinder

Throttle valve

Exhaust

Throttle control

Piston rod

Frame

Guide

Motion

Ram

Upper die

Lower die

Anvil cap

Anvil

Figure 2.5-1 Steam forging hammer.

designed to prevent sticking or fusing of the parts in the dies, improve metal flow, act as a parting agent, and reduce wear of the die. The composition of the lubricant and its impact on the workplace air quality will be discussed later in this section.

A description of a typical forging operation on large parts provides a picture of the exposure of the forge shop personnel. The equipment and personnel for a forging hammer operation are shown in Figure 2.5-3. The forging stock is loaded into a gas or oil fired furnace by the heater and brought to a forging temperature ranging from 430°C (800°F) for aluminum to 1320°C (2400°F) for alloy steel. Box, slot, or rotary hearth furnaces are used to preheat the stock to these forging temperatures.

At the breakdown hammer the surfaces of both dies are blown off with compressed air, and forging compound is applied with spray or swab by the lubricator. The heater and a helper transfer the stock from the furnace to the

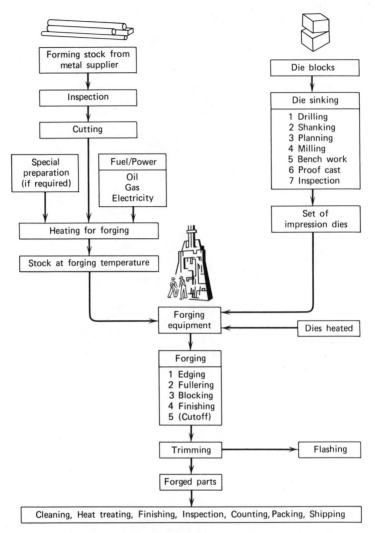

Figure 2.5-2   Hot forging operations. (Courtesy of Forging Industry Association)

breakdown hammer by fork lift truck, by crane, or manually if the piece is small. The piece is positioned on the die by the hammer operator and the lubricator. As many as 8 to 10 blows are struck by the forge hammer operator. Die lubricant is applied to the top die by the lubricator as required between strikes. There is usually some overspray and mist is a significant air contaminant. When an oil-based die lubricant hits the hot workpiece and die, a portion of it burns off, generating a cloud of oil mist, sooty particulates, and a range of gases and vapors of the type noted with poor combustion of a heavy oil. The forging area may be equipped with exhaust hoods of the type shown in Figure 2.5-4. If local exhaust is not available, pedestal fans are positioned as shown in

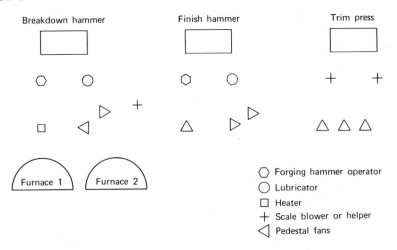

Figure 2.5-3 Hammer operating crew.

Figure 2.5-3 to blow the air contaminants away from the operators. Air contaminants not captured by the exhaust hood eventually are removed by roof exhausters.

When the rough form of the forging is achieved on the breakdown hammer, the forging is returned to the second furnace for reheat to forging temperature. After reheat the forging is transferred to the finish hammer, where the forging

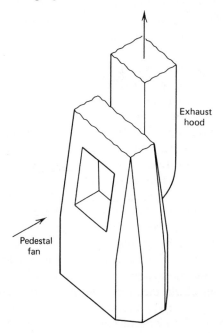

Figure 2.5-4 Local exhaust ventilation on forging hammer.

is completed using the same technique as on the breakdown hammer. Again, forging compound is applied to the dies between strikes and the same air contaminant cloud described on the breakdown hammer is generated. After finish forging, the pieces are taken to the trim press where the metal flashing is removed by a trim die. The flashing is placed in a scrap box and the finishing operations are conducted as shown in Figure 2.5-2.

Forging press operations are similar to hammer operations. In a large press operation, the crew will consist of an operator, barman, oiler, and helper. The operator usually works at a control console some distance from the press while the others work directly at the press and have greater exposure to the air contaminants formed during the operation. Again the stock or billet is first heated in a furnace and then transferred to the press by a fork lift truck. The lubricator blows off the die with compressed air and then sprays on the die lubricant or forging compound. The piece is positioned on the bottom die and is forged with a single stroke. There is exposure to oil mist during application of the forging compound and to the products of combustion during burn-off of the oil.

2.5.2 Air Contaminants

Significant air contamination may occur from furnace operation, the application of die lubricants, the forging operation itself, and the heating of the dies. The various finishing operations will also contribute particulate contamination but since these operations have been discussed in Sections 2.1 and 2.2 they will not be treated in detail in this discussion.

As noted in Figure 2.5-3 stock heating furnaces are positioned adjacent to the forging hammer or press they service. The furnaces are heated electrically or fired with fuel oil or gas. It is not uncommon for the products of combustion to be released from short stacks directly to the workplace and then be removed by roof exhausters. If gas is used, contamination of the workplace with the products of combustion should be evaluated though it is probably not of great significance. However, the release of combustion products from oil fired units may present a significant contamination problem. In addition to the conventional products of combustion, including carbon monoxide, one may have sulfur dioxide formed from fuel oils that contain from 0.5 to 3% sulfur. In most cases, the furnaces used on the forging line are air atmosphere furnaces. Controlled atmosphere furnaces used on special operations may either burn natural gas to provide a reducing or oxidizing atmosphere or, possibly, utilize an ammonia cracking unit to provide the inert atmosphere. The potential health implications of such operations are described in Section 2.8.

A second source of carbon monoxide and other products of combustion is associated with the heating of the dies installed on the forging hammers or presses. Initially dies are brought to temperature in a die furnace and installed hot on the hammer or press. These dies receive supplemental heating to keep

*✱ Big problem
Combustion byproducts are mutagenic*

them at temperature, with an open kerosene or gas burner flame impinging on the die that may contribute to the air contamination occurring at the forging operation.

The application of forging compounds (die lubricants) to hot dies has been a major contributor to forge shop exposures since the industry started. The early lubricants consisted of natural graphite and animal fat added to an oil base. Present die lubricants are a sophisticated blend of components designed for specific forging applications. To be effective, the forging compound must stay on the die face until forging is completed. To accomplish this a variety of additives including coke, sawdust, clay, talc, and asbestos were tried in the industry in the 1940s and 1950s. Compounds of sodium, tin, aluminum, antimony, bismuth, and arsenic were also evaluated as additives by the forging industry during this period. Additives commonly in use at the present time include aluminum stearate, molybdenum disulfide, and lead napthenate. A standard formulation of die lubricant is shown in Table 2.5-1.

Table 2.5-1 *Composition of Forging and Die Lubricant*

Texaforge 7571 (Texaco Corporation)	
25%	Natural graphite
50%	Black oil (crude refined mineral oil)
20%	Petrolatum residuum
5%	Animal oil

Source. Reference 3.

When the hot part is forged, the oil burns off, resulting in a heavy particulate cloud containing oil mist, sooty particulates, trace metals such as vanadium, and polynuclear aromatic hydrocarbons. Sulfur dioxide may also be formed from the combustion of residual oils. In a recent forge shop study, the concentrations of respirable mass particulates and benzo (a) pyrene demonstrated that controls were required (3). In this study, the concentrations of sulfur dioxide, aldehydes, and nitrogen dioxide were not excessive.

Depending on the type of forging, the vehicle used in die lubricants may be one of the following: water, petroleum naptha, light viscosity mineral oil, heavy fuel oils, black oils, or heavy refined cylinder oils. The oils, with the exception of the cylinder oils, have flash points in the range of 149–204°C (300–400°F); the flash point of cylinder oil is 260–343°C (500–650°F). Since the workpiece may be heated to 1320°C (2400°F), one gets ignition of the oils. The particulate contamination from combustion of this oil noted by Goldsmith is shown in Table 2.5-2. The biological effects arising from exposure to oil mist have been studied principally in the United Kingdom (4).

Table 2.5-2 *Particulate Concentrations During Forging*

Particulates	Concentration Range (mg/m^3)	No. of Samples
Total mass	3.2–92.1	43
Respirable mass	3.0–33.3	10
Benzene soluble	1.3–6.0	6
Airborne aromatic	0.3–0.9	6
Benzo(a)pyrene	0.002–0.003	6

Source. Reference 3.

Due to air contamination from the oil based die lubricants, water based forging compounds were introduced in the industry as early as 1950 and are now gaining widespread acceptance when their application is acceptable from a production standpoint (5).

Few data are available on the concentrations of metal dust or fumes generated from the workpiece during forging operations. The contamination generation rate depends on forging temperature, vapor pressure of the metal at forging temperature, metal flow characteristics, and the oxidizing potential of the metal. If conventional steels are worked, the concentration of iron in the air is not significant; however, the forging of alloys containing chromium and nickel may warrant air sampling. Specialty forging of highly toxic metals may contribute significant air contamination unless specific controls are installed to control air contaminants.

In the past asbestos was used to cover large workpieces during forging to reduce heat loss, but now fiberglass blankets may be used. During forging, the cloth or blanket is shredded and may become airborne. When the finished part is cleaned by abrasive blasting, the fiberglass or other fibers adhering to the metal will become airborne also.

The machining of die blocks may be conducted as a part of the forge operation or done by an outside supplier. The operations involved in "cutting" or "sinking" the die cavity are listed in Figure 2.5-2. In most cases the metal stock is a steel alloy that may contain chromium and molybdenum. Since most of the work is milling using coolants, it does not represent a significant source of air contamination. Specific comments on machining operations are presented in Section 2.10.

2.5.3 Heat Stress

In high production forge shops, heat stress may be a major health problem. The principal heat load is due to radiation from furnaces, dies, and the workpiece and metabolic load when material handling is done manually. In certain geographical areas, convective heat load may be significant. A survey must be conducted to determine the origin of the heat load to the worker so that controls may be installed. Controls may include reduction of the physical work load by improved materials handling, shielding of furnaces to reduce the radiation load, and spot cooling of the worker. Impressive reduction in heat stress can be accomplished if reflective shields are installed on the stock heating furnaces although one must face the challenge of keeping these shields clean to maintain maximum effectiveness. The opening height of furnace doors or slots must be kept at a minimum to reduce exposure of the heater to direct furnace radiation. The industry frequently uses steam and airlines mounted at the bottom of the door that release steam or air when the door opens. The steam release may reduce the radiant load somewhat although no data are available to support this claim. Lightweight movable plates or heat shields in front of the furnace loading door may also be effective if used efficiently by the heater. Pedestal fans are commonly used for man cooling and to blow smoke away from the forging crew. Obviously, if the air temperature is below 35°C (95°F) this is effective; it may not be useful if air temperatures exceed the nominal skin temperature of 35°C (95°F) as this could contribute to the worker's convective heat load.

2.5.4 Noise

Forging was one of the first industrial operations to be identified as causing noise-induced hearing loss. The noise hazard is due to hammer impact noise (e.g., a 3–4 ton ram dropping on a 20 ton anvil) and, to a lesser degree, "pink" noise from the operation of steam and air cylinders and pneumatic solenoids. The hammer operation may be as slow as one blow per minute or as high as 60 blows per minute for a small airlift hammer. Impact noise may be as high as 140 dBA. The air release noise can be controlled by mufflers; however, reduction of the primary forging impact noise is more elusive. This forge shop noise problem normally is controlled by ear protection and a complete hearing conservation program. It may be possible to partially isolate the forge hammer or press with partitions or lead-lined curtains to minimize the exposure of other workers.

2.5.5 Eye Hazard

The periodic viewing of the hot workpiece and the furnace interior presents a possible infrared radiation exposure that has not been fully evaluated. Heat cataracts may develop depending on radiation intensity, time exposed, and age of the worker. The information contained in Section 3.17 on glass production also applies to the forging industry.

2.5.6 Control

A combination of local and dilution exhaust ventilation is used in most large forge shops to control air contaminants. The small job shop frequently operates without local exhaust ventilation control. Dilution exhaust ventilation is reasonably effective for carbon monoxide control from many fugitive sources but does not control the smoke cloud from forging operations. Hot forging hammers and presses utilizing oil-based die lubricants should be equipped with local exhaust ventilation. The industry frequently installs local hooding, which also acts as a shield to control flying scale during the air blowoff of dies. To date there are no design criteria published on such local exhaust ventilation. Air contaminants arising from stock heating furnaces may be eliminated if stock can be heated by electrical resistance or induction heating methods.

Protective clothing requirements vary greatly with the type of forging. One should at least consider head protection and the mandatory use of protective goggles and shoes. On certain operations gauntlet gloves and shoulder length fireproof sleeves are necessary. If there is heavy use of oil-based die lubricants, suitable oil- and fireproof aprons and leggings should be worn by the forging crew. For the heaters, goggles with colored lenses and wire mesh face screen for infrared protection may be in order. Employees exposed to excessive noise levels should use hearing protection.

REFERENCES

*1   J. Jenson, Ed., *Forging Industry Handbook,* Forging Industry Association, Cleveland, OH, 1970.

2   T. Lyman, Ed., *Metals Handbook,* 8th ed., Vol.5, American Society for Metals, Metals Park, OH, 1964.

3   A. H. Goldsmith, K. W. Vorpahl, K. A. French, P. T. Jordan, and N. B. Jurinski, *Am. Ind. Hyg. Assoc. J.,* **37,** 217 (1976).

4   H. A. Waldron, *J. Soc. Occup. Med.,* **27,** 47 (1977).

5   J. W. Engelhardt, *Am. Mach. 21,* 60 (Jan. 1975).

2.6 FOUNDRY OPERATIONS

Metal founding or casting is the pouring of molten metal into a cavity formed in some type of molding media. The principal media used is silica sand. The mold cavity may contain a refractory core to define a void in the casting. After cooling, the mold is taken to a shakeout facility where the molding media is removed from around the casting. The casting is cleaned, extraneous cast metal is removed, and the molding sand is recycled to be conditioned for reuse. These foundry processes are shown in Figure 2.6-1 for a gray iron foundry using a cupola furnace (1).

The potential health hazards in foundry operations are exposure to various air contaminants and physical conditions including noise, heat, and vibration (2, 3). A survey of 281 foundries in Michigan in 1969 revealed that more than 7% of the employees were exposed to serious health hazards (4). It is reasonable to presume that conditions are at least that serious in other states.

This section discusses the occupational health hazards in molding, coremaking, casting, shakeout, and finishing. Limited attention is given the

* = Silica, CO, Lead, (CHO, Phenol - from cores)   ± Cd, Berylium

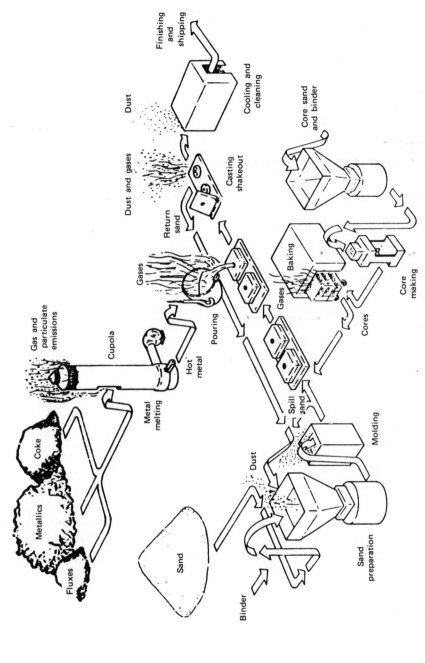

Figure 2.6-1 Gray iron cupola foundry. (From Reference 1)

49

cleaning operations since they are covered in Section 2.1. The discussion focuses on the casting of iron and steel; however, specific problems associated with aluminum, magnesium, brass, and bronze foundries are identified.

2.6.1 Molding

The majority of both ferrous and nonferrous castings are produced in green sand molds, that is, with sand that is damp and plastic. This molding sand is a mix of silica sand with bentonite or kaolinite clay (fire clay) as a binder, various cereal or wood flours, and water or oil. In many foundries finely ground soft coal (sea coal) is added to the molding sand. When the metal is poured, the sea coal pyrolyzes forming gases that reduce metal penetration into the sand and improve the metal surface.

The silica sand is mixed with the other ingredients in a muller. Loading of the high silica sand is dusty so the muller must be provided with local exhaust ventilation (5, 6). After mixing, the sand is placed in storage bins or silos and when required it is mechanically transferred to delivery bins at the molding positions. At this time the sand is a wet cohesive mass and it is not a dust hazard. After the casting and shakeout operations the friable, dusty sand is conditioned so that it can be reused. The potential hazard from the molding sand varies as it is processed through the foundry as shown in Table 2.6-1 (2).

Table 2.6-1 *Factors Influencing Hazards from Foundry Sand*

Factors	Effect on Health
Physical states of sand	
New sand, screened and dust-free	Limited health hazard
New wet sand	Limited health hazard
New sand, fine and dry	Potential hazard
Used sand	Potential hazard
Chemical composition of sand	
Olivine (mixture of silicates and oxides)	Limited health hazard
Clays (mixture of silicates) and inert binders (dextrose)	Limited health hazard
Sand contaminated with toxic metallic particles	Potential hazard
Pure quartz (silica, SiO_2)	Potential hazard
Method of handling	
Unloaded by vacuum system	Potential hazard
Compressed air in closed pipe system	Potential hazard
Belt conveyor	Potential hazard
Clamshell	Potential hazard
Mechanical shovels	Potential hazard
Work done outdoors	Potential hazard
Manual indoor work (sweeping, sand cutting)	Potential hazard

Source.   Reference 2.

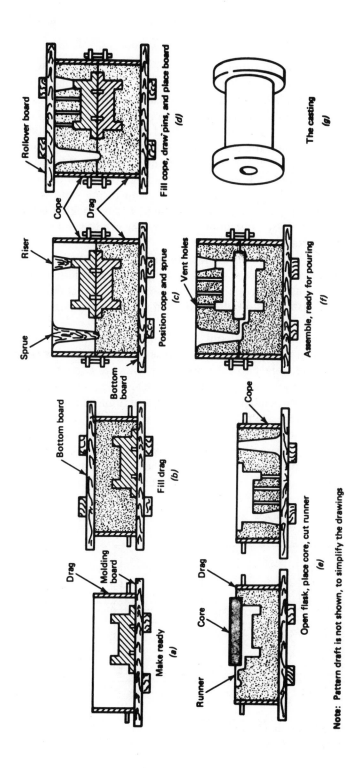

Figure 2.6-2 Green sand molding. (From *Manufacturing Processes* by A. Roberts and S. Lapidge. Copyright 1977. Reproduced with permission of McGraw-Hill Book Co.)

Rollover board

Fill cope, draw pins, and place board
(d)

The casting
(g)

Cope

Drag

Riser

Sprue

Position cope and sprue
(c)

Vent holes

Assemble, ready for pouring
(f)

Bottom
board

Bottom board

Fill drag
(b)

Cope

Open flask, place core, cut runner
(e)

Drag

Molding
board

Make ready
(a)

Core

Drag

Runner

Note: Pattern draft is not shown, to simplify the drawings

51

Green sand molding

The steps involved in making a simple mold from green sand are shown in Figure 2.6-2. An image or pattern of the metal part to be cast is made from wood or metal. The pattern is designed so that it can be withdrawn after the sand has been packed around it. In the manual bench molding operation, a metal or wooden frame called a flask is used to hold the mold. The top half of the flask is called the cope, and the bottom is the drag. If additional height is needed, an extra section known as the cheek is mounted between the cope and the drag. The drag is inverted and placed on a molding board; the pattern is positioned on the board inside the flask. A fine, clean facing sand is then screened (riddled) over the pattern and board. Next, heavy backing sand is rammed in place manually or by a pneumatic tamper. The excess sand is struck off and the drag is inverted.

A parting compound is dusted on the face of the pattern so that later the completed cope can be easily removed. The matching cope is positioned on the drag and pinned in place and the cope pattern is assembled. The cope is then completed in the same fashion as the drag; wooden spacers are positioned to form the sprue and riser channels.

The sprue and associated gating convey the molten metal to the mold cavity. The riser acts as a vent and provides a reservoir of molten metal to handle shrinkage when the metal cools. Various vent holes are also introduced into the mold. The cope is then removed and the cope and drag are positioned face up so the patterns can be removed. Sprue, riser, and runner forms are also removed at this time. If there are to be channels or voids in the casting, sand cores must be positioned in the mold. The making of the sand cores will be described later; however, these are refractory parts with mechanical strength and are positioned in core prints in the mold. With the cavity defined, the cope and drag are reassembled and clamped together ready for pouring.

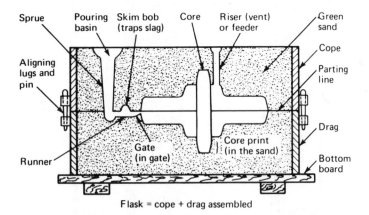

Flask = cope + drag assembled

Figure 2.6-3 Completed sand mold. (From *Manufacturing Processes* by A. Roberts and S. Lapidge. Copyright 1977. Reproduced with permission of McGraw-Hill Book Co.)

The finished sand mold is shown in figure 2.6-3.

As indicated earlier, the molding sand is not dusty and does not result in a significant silica exposure during the molding operation, however, parting material may represent a dust exposure.

The most hazardous parting compound is silica flour, which is dusted on the mold with a small bag. Although used widely at one time, silica flour has for the most part been replaced by nonsiliceous materials. The common parting compounds are shown in Table 2.6-2 (3). If the materials are applied wet they may be suspensions in water, aliphatic hydrocarbons, 1,1,1 trichloroethane, or a Freon ® solvent. Flammable solvents should not be applied on hot patterns. If spray application of the suspensions is required, either airless or electrostatic spray techniques should be used to minimize overspray and rebound.

Table 2.6-2 *Foundry Parting Compounds*

Dry Release Agents	Wet Release Agents
Graphite	Dry release agents in suspension
Mica	Bentonite suspension
Metal stearates	Mineral oil
Molybdenum disulfide	Fatty acids (oleic acid)
Polyethylene	Silicones
Polyvinyl alcohol	Soya lecithin
Silica flour	
Stearic acid	
Talc	

Source. Reference 3.

In a high production shop the mold described earlier would be made on molding machines. This equipment is designed to pack the sand firmly in the flasks and manipulate the flasks, pattern, and completed molds semiautomatically. The molding machines' tables are equipped with air cylinders that initially jolt the sand into place in the flask and then squeeze the sand to pack it around the pattern in a reproducible manner. If the work is too large for machine molding, it is done with large floor molds using flasks that are handled by crane. These flasks are filled with molding sand using a mobile sand slinger, which is a centrifugal blower located at the end of a boom and supplied with molding sand from a hopper. The operator can remotely direct the high velocity sand stream to various parts of the flask and produce a compact mold. If the cast part is too large for available flasks, a pit is dug in the foundry floor and the molding and pouring is conducted in place.

In addition to the conventional green sand molding processes described in this section, there are a number of specialty molding processes used principally in steel and nonferrous foundries that are described in the following sections.

Shell molding

The shell molding process is used for the high production of complex-shaped, small parts. The sand for shell molding is coated with phenol-or urea-formaldehyde resin using a cold, warm, or hot coating technique. In the cold technique the resin with hexamethylene tetramine and calcium stearate is blended with dry sand in a mixer. In the warm coating process the resin-sand mix is dried with a 150–180°C (300–350°F) air stream. Flaked resin is blended with the sand in the hot coating technique, quenched with a slurry of hexamethylene tetramine and calcium stearate, and then aerated at ambient temperature. It is common practice to obtain precoated shell sand, thereby eliminating these processes at the foundry.

A manual, dump-box technique for shell molding is shown in Figure 2.6-4. The metal match plate is the pattern in this process reflecting the geometry of the part to be cast. The pattern is sprayed with a parting agent such as a silicone, mounted on the resin-sand dump box, heated by the oven, and the dump box is inverted. The resin hits the hot pattern and a skin of partially cured resin-sand is formed over the pattern. The dump box is turned upright and the oven is positioned over the pattern to effect a complete cure at temperatures of 316–427°C (600–800°F). The cured shell is stripped from the pattern and assembled with a matching shell to form the mold. The cope and drag are either glued or clamped together or, if small, they are supported by sand or steel shot in a flask for casting.

The layout for a semiautomatic dump-box shell molding machine is shown in Figure 2.6-5. The operator removes a completed cope from the machine and places it on the rack. He or she then removes the matching part of the shell mold (drag) and places it on the glue applicator and assembles the cope and drag on the mold press. The completed mold is then placed on a storage rack, and the process is repeated.

The operator of a shell molding machine is exposed to phenol, formaldehyde, ammonia, hexamethylene tetramine, and carbon monoxide. In a study of a well-exhausted shell molding machine, carbon monoxide concentrations ranged from 15–20 ppm, ammonia from 3–10 ppm, phenol and formaldehyde were not detected, and respirable dust concentrations ranged from 0.82 to 1.4 mg/m³ (7).

Local exhaust ventilation is required on shell molding machines at the cope hanger, mold press, and storage rack. A fresh air supply diffuser may be required to control heat stress on this operation (7).

Investment casting

The lost wax or investment casting method is an ancient technique that is now used industrially for precision casting of a range of products including turbine vanes and nozzles. As shown in Figure 2.6-6, a metal die is used to cast expendable patterns made from hard wax or plastic such as polystyrene. The formed patterns are assembled in a cluster with a common sprue so that a large number of parts can be obtained from one casting. The cluster or tree is then placed in a metal flask and invested with a mixture of plaster of paris, talc, silica, and water (or possibly an alcohol). After the investment is cured, the

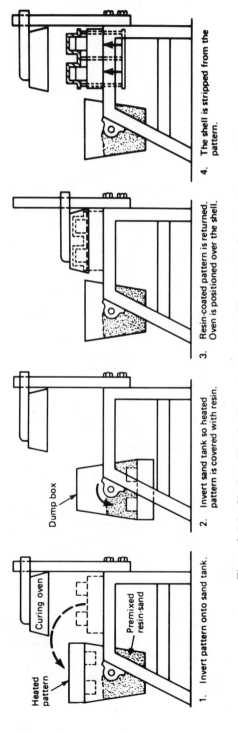

Figure 2.6-4 Shell molding by dump-box method. (From *Manufacturing Processes* by A. Roberts and S. Lapidge. Copyright 1977. Reproduced with permission of McGraw-Hill Book Co.)

1. Invert pattern onto sand tank.

Heated pattern

Curing oven

Premixed resin-sand

2. Invert sand tank so heated pattern is covered with resin.

Dump box

3. Resin-coated pattern is returned. Oven is positioned over the shell.

4. The shell is stripped from the pattern.

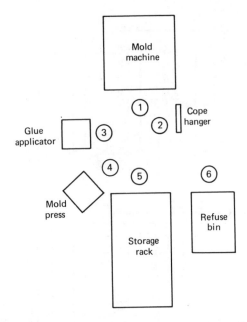

Figure 2.6-5 Plan view of a shell molding station. Numbers in circles indicate operator positions. (From Reference 7)

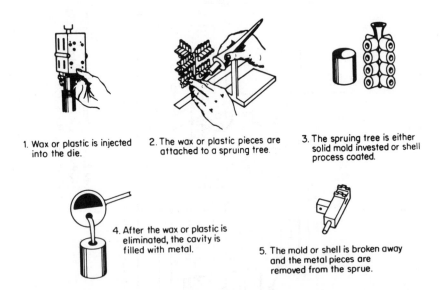

1. Wax or plastic is injected into the die.

2. The wax or plastic pieces are attached to a spruing tree.

3. The spruing tree is either solid mold invested or shell process coated.

4. After the wax or plastic is eliminated, the cavity is filled with metal.

5. The mold or shell is broken away and the metal pieces are removed from the sprue.

Figure 2.6-6 Investment casting. (From *Manufacturing Processes and Materials for Engineers* by L. E. Doyle. Copyright 1969. Reproduced with permission of Prentice-Hall, Inc.)

flask is inverted and placed in an oven. The wax patterns melt and the wax drains out of the flask; if polystyrene is used, it is vaporized. Molten metal is poured into the cavity, the metal is cooled, and the parts are retrieved.

The hazards in this operation vary depending on the exact materials in use. The preparation of the silica slurry involves an exposure to silica sand and alcohol. Low temperature melting of the wax should not present a problem unless chlorinated waxes are used in the process. The vaporization of the polystyrene patterns may release a number of thermal degradation products depending on the oven temperature. Removing the parts may involve an abrasive blasting operation with possible dust exposure.

Full mold

In this process an expendable pattern in the image of the part to be cast is made of polystyrene. In most cases the pattern can be constructed so that cores are not required. The pattern is positioned in a molding box with the necessary sprues and risers also made of polystyrene. Molding sand is placed around the pattern and gently rammed into place. When the molten metal is poured into the sprue, the polystyrene vaporizes and the gases and vapors diffuse into the sand mass. The entire polystyrene pattern vaporizes in this manner and the molten metal fills the space originally occupied by the pattern. An advantage of this process is the minimal cleaning required of the cast part.

The possible hazards from this operation rest solely with the venting of gases and vapors from the vaporized polystyrene. Laboratory investigations of plastic thermal degradation suggest that one should be prepared to sample for styrene monomer, carbon monoxide, and possibly benzene.

2.6.2 Coremaking

The core is a refractory element placed in the mold to define a cavity in the final casting. Since the molten metal will flow around it, the core must be mechanically strong at that point and yet become friable after pouring and cooling to allow easy removal from the casting. Cores are made in a fashion similar to that for the molds described previously. In the oldest system in use, core sand is prepared in a muller by mixing silica sand with an organic binder such as linseed oil and starch or dextrine. The sand is packed in a core box with a cavity defining the shape of the core. The fragile core is then removed and cured in a core oven at approximately 200°C (400°F).

Core ovens handling oil-based cores are notorious air pollution sources releasing acrolein and other aldehydes to the workplace and the neighborhood. In the past decade there has been a revolution in core manufacturing techniques. The new coremaking procedures involve a series of binder systems (3). Certain binder systems require oven heating, others require gassing to cure the system, and there are many no-bake systems. As described later, these systems may or may not release air contaminants during curing, but most do in pouring and shakeout operations.

In high production foundries, cores are made on automatic core machines with multiple cavities. The core machines fill the cavities with core sand (coreblowing), gas, purge, and eject the core. After the cores are ejected, the operator cleans and repairs them and applies a coating. The cleaning process may involve a simple dressing-off with a knife and hand or machine sanding with some dust exposure. The repair of small defects is done with a putty knife and a filler compound. The cores may be dip or spray coated; a popular coating is a paraffin material in an isopropyl alcohol vehicle. The entire operation may be conducted by one person, or in a large foundry the tasks may be divided between a muller operator, core machine operator, and a core finisher.

The specific resin systems presently in use and worker exposures to airborne contaminants are outlined in the following sections.

Sodium silicate system

This system has been in widespread use since its introduction in the mid 1950s; it represents a minimal health hazard to the workers. The core sand is prepared by mixing 3–6 parts of sodium silicate (water glass) to 100 parts silica sand with certain other additives. After the core box is filled and faced, the core is cured by passing carbon dioxide through the core sand. A reaction takes place forming sodium carbonate and a silicon dioxide gel and fixing the sand in place with a strong bond. Chemical activators now used in lieu of carbon dioxide include ferrosilicon, sodium silicon fluoride, dicalcium silicate, Portland cement, glycerol triacetate, glycol diacetate, and diacetin (3).

Workers handling the concentrated water glass, a strongly alkaline solution, should wear personal protective equipment including gloves, aprons, and goggles. The chemical activators also require care in handling.

The carbon dioxide vented from the core box does not normally present a problem in an open workplace; however, if the process is conducted in a molding pit or other enclosed space, significant carbon dioxide concentrations may exist. If carbon dioxide is used to cure the cores, the thermal degradation products released during the later pouring and shakeout operations are those common to any sand molding operation. The use of the organic additives will result in the formation of carbon monoxide and other products during pouring. A complete description of the thermal degradation products of these activators is not available at this time.

Hot box system

In this process thermosetting resins such as phenol-formaldehyde, urea-formaldehyde, furfuryl alcohol-formaldehyde, or other combinations are mixed with sand and a catalyst to form a system that will cure to a solid mass in a heated core box (3). A common blend of components is 100 parts of sand, 2 parts of the resin, and 0.4 parts of the catalyst. A variety of catalysts including ammonium chloride and ammonium nitrate are in use. After mixing in a muller, the sand, resin, and catalyst are injected into a mold heated to a temperature of 204–260°C (400–500°F). The compressed core sand mix is cured to a solid mass in 1 or 2 min. The core is then ejected to a cooling station where the cure continues throughout the mass of the core. During this period, significant off-gassing occurs and good local exhaust ventilation is required (7).

Precautions are required in handling the resin and catalyst in concentrated form while preparing the mix. Such precautions should include skin and eye protection for both the urea- and phenol-based resins. Ventilation control is required on the mixer, the coremaking machine including the coreblowers, the cool-down location, and the pouring, casting, cooling, and shakeout stations.

The air contaminants released during hot box core manufacture include ammonia, formaldehyde, phenol, and furfuryl alcohol. The exact composition depends on the type of resin system in use. Air concentration data on operations with good control have been developed in a recent NIOSH study (7).

No-Bake System

Efforts to eliminate oven and core box heating processes led to the development of a series of sand-resin-catalyst systems that cure at room temperature. As will be noted later, these systems do involve unique exposure patterns to contaminants not previously seen in the foundry environment.

The most common system is a phenol resin used with or without furan resins. Acid catalysts including phosphoric, toluene sulfonic, benzene sulfonic, and sulfuric acids are used with these systems.

Urethane no-bake systems are formed when phenol resin is used in conjunction with isocyanates. This system requires such catalysts as cobalt napthenate, triethylamine, and dimethyl ethylamine. Epoxy resin systems use amine hardeners or catalysts that are skin sensitizers and possibly pulmonary sensitizers.

A range of hazards exist from no-bake resin operations. Skin and eye contact with the principal resins should be avoided. The strong acids used as catalysts in the phenolic-furan systems demand attention to safe handling procedures and personal protective equipment. Exposure to the isocyanates may cause pulmonary asthma at low concentrations and the amine hardeners used in the epoxy systems are known skin sensitizers and potential respiratory sensitizers. Local exhaust ventilation is required where the sand-resin systems are weighed out and mixed, at the core machine and core hanger storage, and during pouring and shakeout. Skin contact must be minimized and scrupulous housekeeping practices followed and encouraged by disposable paper covers on work surfaces. A set of tools should be assigned solely to the resin-catalyst work stations to prevent contamination of other work sites.

Shell coremaking system

This system is identical to the shell molding process described in Section 2.6.1 and includes skin contact with hexamethylene tetramine and air contaminants such as ammonia, phenol, and formaldehyde. The thermal degradation products noted in pouring and shakeout are similar to those for other phenol- and urea-based systems.

2.6.3   Metal Casting

The process of casting includes preparation of the charge materials, preheating of the charge furnace and ladles, melting of the charge in the furnace, fluxing of the melt both in the furnace and at the ladle to remove silicates and oxides, inoculation of the charge with materials for improved metallurgical properties, tapping of the furnace, pouring from furnace to a receiving ladle, and subsequent transfer to smaller pouring ladles and pouring of the melt into the prepared molds. The major health hazards in both ferrous and nonferrous operations include exposure to toxic metal fumes, carbon monoxide and other toxic gases, and heat stress. These exposures are discussed in the following major categories.

The potential exposures to metal fumes are listed in Table 2.6-3 and to thermal degradation products from the molds and cores in Table 2.6-4.

Table 2.6-3   *Dust and Fume Exposures from Metal Melting and Pouring*

Metal	Dust or Fume	Occurrence
Iron and steel	Iron oxide	Common
	Lead, Leaded steel	Common
	Tellurium	Rare
	Silica	Common
	Carbon monoxide	Common
	Acrolein	Rare
Bronze and brass	Copper	Common
	Zinc	Common
	Lead	Common
	Manganese	Rare
	Phosphine	Rare
	Silica	Common
	Carbon monoxide	Common
Aluminum	Aluminum	Common
Magnesium	Magnesium	Common
	Fluorides	Common
	Sulfur dioxide	Common
Zinc	Zinc	Common

Table 2.6-3 *Dust and Fume Exposures from Metal Melting and Pouring—Continued*

Metal	Dust or Fume	Occurrence
Cadmium	Cadmium	Common
Lead alloys	Lead	Common
	Antimony	Common
	Tin	Common
Beryllium	Beryllium	Commom
Beryllium–copper	Beryllium	Common
Uranium	Uranium	Rare

Source. Reference 2.

Table 2.6-4 *Potential Hazard Evaluation of Chemical Emissions from Foundry Molds*

	Green Sand	Sodium Silicate Ester	Core Oil	Alkyd Isocyanate	Phenolic Urethane	Phenolic No-Bake	Low N$_2$ Furan- H$_3$PO$_4$	Med N$_2$ Furan- TSA	Hot Box Furan	Phenolic Hot Box	Shell
Carbon monoxide[a] (30 min)	A[b]	A	A	A	A	A	A	A	A	A	A
Carbon dioxide[a] (30 min)	B[c]	B	B	B	B	B	B	B	B	B	B
Sulfur dioxide	B	C[d]	C	C	C	A	B	B	C	C	B
Hydrogen sulfide	B	C	C	C	C	B	B	B	C	C	C
Phenols	B	C	C	C	B	B	C	C	C	C	C
Benzene	B	B	B	B	B	B	C	B	B	B	B
Toluene	B	C	C	B	B	B	C	B	C	C	B
meta-xylene	C	C	C	B	B	C	B	B	C	C	C
o-Xylene	C	C	C	B	B	C	B	C	C	C	C
Naphthalene	C	C	C	C	C	C	C	C	C	C	C
Formaldehyde	C	C	C	C	C	C	C	C	C	C	C
Acrolein	C	C	C	C	C	C	C	C	C	C	C
Total aldehydes (acetaldehyde)	C	C	C	B	C	B	C	B	C	C	C
Nitrogen oxides	B	C	C	B	C	C	C	B	B	B	C
Hydrogen cyanide	C	C	C	B	B	C	B	B	B	B	B
Ammonia	C	C	C	C	C	C	C	B	A	A	B
Total amines (as aniline)	C	C	C	C	B	C	C	B	B	B	B

Source: Reference 8.

[a] Measured in ppm, determined from graphical integration.

[b] A = chemical agent present in sufficient quantities to be considered a definite health hazard. Periodic monitoring of concentration levels in work place recommended.

[c] B = chemical agent present in measurable quantities, considered to be a possible health hazard. Evaluation of hazard should be determined for given operation.

[d] C = chemical agent found in minute quantities. Not considered a health hazard under normal conditions of use.

Furnace melting

Open hearth (Figure 2.6-7)

Historically, large steel foundries have utilized open-hearth furnaces to produce both ingot and steel castings. The units, ranging in melt size from 10 to 600 tons, are charged with 35–60% scrap metal with a balance of pig iron. Acidic metal oxides are formed in the melt. Limestone and dolomite are added to the charge as fluxing agents. The metal oxides react with these fluxing materials and the basic furnace lining to form a slag that is removed before pouring.

The major health hazards from open-hearth operations are heat stress and carbon monoxide exposure during maintenance operations. The furnaces are also a major air pollution source of metal fume. Open-hearth furnaces are still used in the steel industry but their application in steel foundries has decreased rapidly in the last 20 years and there are now probably less than a dozen operational facilities in the United States.

Arc furnaces (Figure 2.6-8)

The electric arc furnace, handling 2–200 tons per melt, has replaced the open hearth as the major furnace in large steel foundries. The furnace is charged with ingot, scrap, and the necessary alloying metals. An arc is drawn between the three carbon electrodes and the charge, heating the charge and quickly melting it. A slag cover is formed with various fluxing agents to reduce oxidation of

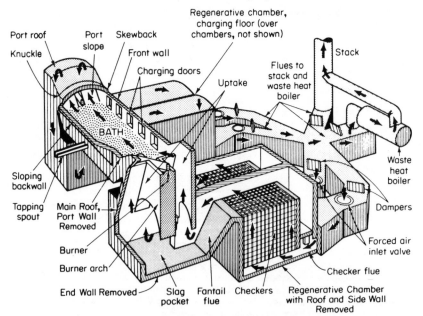

Figure 2.6-7 Open hearth furnace. (From *Manufacturing Processes and Materials for Engineers* by L. E. Doyle. Copyright 1969. Reproduced with permission of Prentice-Hall, Inc.)

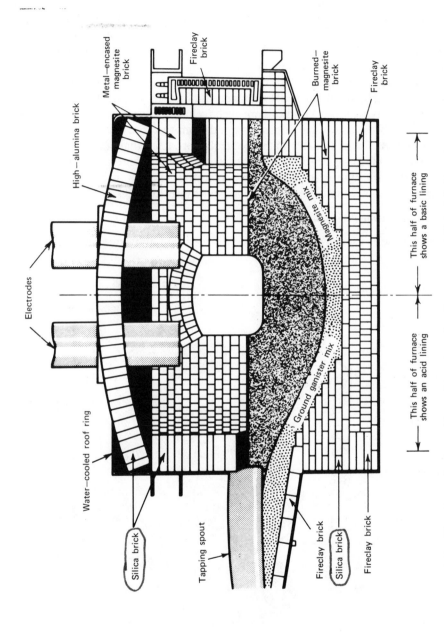

Figure 2.6-8 Electric arc furnace. (From *Manufacturing Processes and Materials for Engineers* by L. E. Doyle. Copyright 1969. Reproduced with permission of Prentice-Hall. Inc.)

Electrodes

Water—cooled roof ring

Silica brick

Tapping spout

Fireclay brick

Silica brick

Fireclay brick

High—alumina brick

Metal—encased magnesite brick

Fireclay brick

Burned—magnesite brick

Fireclay brick

Magnesite mix

Ground ganister mix

This half of furnace shows an acid lining

This half of furnace shows a basic lining

surface metal, refine the metal, and protect the roof of the furnace from damage from excessive heat radiation. When the melt is ready for pouring, the electrodes are raised and the furnace is tilted to pour into a receiving ladle.

These furnaces produce tremendous quantities of metal fume resulting in both a workplace and an air pollution problem. Local exhaust ventilation systems based on side draft or enclosing hoods are available to handle fume generation during the melt cycle (6). A slag door hood and pour spout hood must be included in the ventilation system. Since metal fume will escape during the operation, high volume roof exhausters are located over the furnace area to remove fugitive losses. Another approach to local exhaust ventilation on this equipment is the use of partitions dropped from the roof to form a modified canopy hood. The problem with this control approach is that one is faced with a difficult air cleaning problem, that is, a large air volume and a low fume concentration. One also encounters a serious noise problem from the intermittent make and break of the arc on these furnaces.

Induction furnace (Figure 2.6-9)

This furnace is used widely in both nonferrous and alloy steels foundries. A

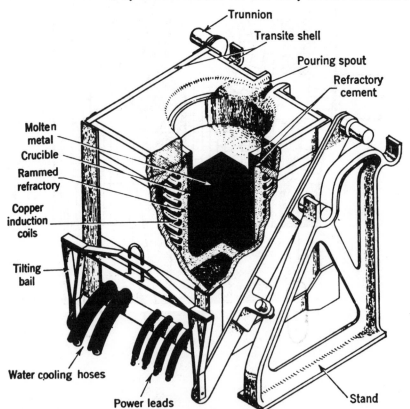

Figure 2.6-9 Induction furnace. (From *Manufacturing Processes and Materials for Engineers* by L. E. Doyle. Copyright 1969. Reproduced with permission of Prentice-Hall, Inc.)

melting refractory is surrounded by water cooled copper coils powered by a high frequency power supply. The outer winding induces current flow in the outer edge of the metal charge and, due to the high resistance, the metal charge is heated and melting progresses from the edge of the charge to the center.

The metal fumes are best controlled by enclosing hoods although canopy hoods and dilution ventilation are utilized frequently (5). For low alloy steels, the canopy hood or dilution ventilation may provide adequate control; however, an enclosing hood is required for nonferrous metals.

Crucible furnaces (Figure 2.6-10)   Crucible made from non-ferrous metal.

The crucible containing the charge is heated directly by a gas or oil burner or, occasionally, coke. This furnace is widely used for nonferrous alloys. The principal hazards are carbon monoxide, metal fumes, burner noise, and heat stress. Ventilation is usually accomplished by a canopy hood at the furnace line although perimeter slot hoods provide better control.

Cupola (Figure 2.6-11)

Over 60% of the gray iron castings produced in the United States are produced by the cupola. It is the most economical way to convert scrap and pig iron to usable molten iron. As produced, the metal is gray iron; if inoculated with magnesium or cerium at the ladle, a ductile iron is formed. As shown in Figure 2.6-11, alternate layers of coke, limestone, and metal are periodically charged to the furnace.

A major hazard from cupola operation is exposure to carbon monoxide. This is especially true if twin cupolas are in use and one is under repair during (reflux) operation of the second. Iron oxide fume is not a direct problem in the

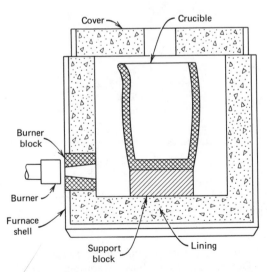

Figure 2.6-10 Crucible furnace. (From *Manufacturing Processes and Materials for Engineers* by L. E. Doyle. Copyright 1969. Reproduced with permission of Prentice-Hall, Inc.)

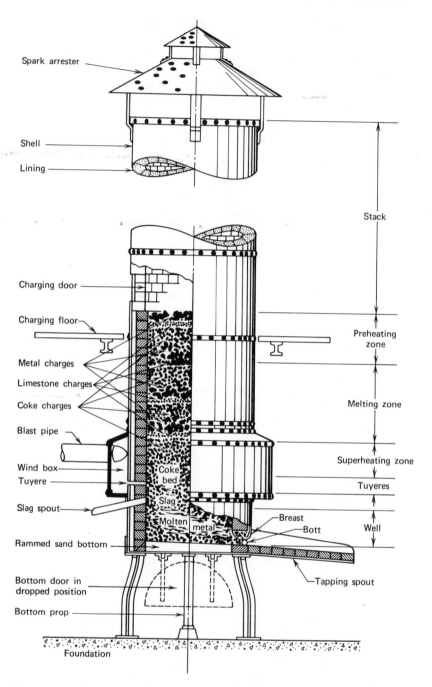

Figure 2.6-11 Cupola furnace. (From *Manufacturing Processes and Materials for Engineers* by L. E. Doyle. Copyright 1969. Reproduced with permission of Prentice-Hall, Inc.)

workplace, but scrap iron may contain small quantities of toxic metals such as lead and cadmium that may present significant exposures.

Personnel who operate and repair cupolas have a serious heat problem. Where a fixed pouring station is used, an air conditioned work station may be installed for the ladle man. Reflective barriers are also effective in minimizing the radiant heat load.

Transfer, pouring, cooling

Fume releases during pouring to the receiving ladle, subsequent transfer to the pouring ladle, and pouring at the mold line require local exhaust ventilation. Two approaches are possible for ventilation control of pouring operations. If the molds are transferred by conveyor, the pouring may be done at a fixed location with an installed local exhaust hood. If the molds are poured on the floor, the ladle must be equipped with a mobile hood.

Inoculation of heats with special metals does, however, place the worker in close proximity to the furnace or ladle with resulting high exposures to heat, metal fume, noise, and accidental metal splashes. The most recent inoculation techniques including bell and sandwich inoculation reduce these hazards although they still warrant attention (7). Lead, a common inoculant to improve machinability of steel, illustrates the potential hazard during inoculation. Depending on how it is added, from 20 to 90% of the lead may be released to the atmosphere. Concentrations of lead up to 1.0 mg/m³ may be noted at the operator's level during inoculation and high levels may exist in the crane cab over the ladle area.

Silicate forming materials and oxides must be removed from the metal before pouring. This requires the addition of fluxing agents and removal of the slag that is formed. To protect the worker from fume exposure and heat stress, a combination of shielding and distance is frequently used. On large operations with severe heat stress, mechanical slagging is conducted by remote control.

Operators directly involved in handling the molten metal may require personal protective equipment including aluminized clothing, tinted glasses and faceshields, and head protection. A combination of engineering and work practice controls must be installed to handle heat stress. The acclimatization of the worker requires at least 1 week so care should be observed during the first days of the warm season and during the start-up week after vacation.

The crane operator has a special exposure that warrants attention. The metal fume lost from the floor operations is cleared by roof exhausters and high air concentrations of fume may be noted at the crane level. Obviously, if remote radio controlled cranes are used, the operator is not exposed. If an operator must be positioned in the crane, one should consider enclosure of the cab with a supply of filtered, conditioned air.

After pouring and before shakeout, molds are moved by a conveyor to a staging area for cooling. A critical exposure to carbon monoxide may occur in this area, especially if sea coal is added to the molding sand to improve the surface of the casting. The sea coal forms methane, which burns off with the

in staging area

production of carbon monoxide. The maximum concentration is noted directly after pouring with an exponential decay with time. The author has noted concentrations of carbon monoxide of 200–300 ppm in the staging area of a small gray iron foundry. Control may be achieved by routing pallet molds through a ventilated cooling tunnel.

Foundries casting very large parts with limited melt capacity use hot topping powders to decrease the rate of cooling in the feeding head. These exothermic powders may be based on aluminum with an oxidizing agent such as potassium or sodium nitrate and a fluoride of an alkali metal that controls the reaction. The hot topping powders may reach temperatures of 1800°C (3270°F) with the release of a heavy particulate cloud. This operation has not been thoroughly studied and the chemical species released will vary widely depending on the exact composition of the powder.

2.6.4 Shakeout

The casting is removed from the mold at the shakeout position. Small to medium-sized castings are placed on a vibrating screen; the molding sand drops to a hopper through the screen for return by conveyor for reconditioning; the flask is routed back to the molding line and the casting is hooked free for cleaning. Small castings can be removed from the flasks by a "punch-out" process. This method is superior to shakeout since it generates less dust. Shakeout is still required for removal of surface sand and cores. Shakeout is a hot, demanding job with serious exposure to particulates, gases, vapors, noise, and vibration.

Considerable data are available demonstrating significant air concentrations of silica at foundry shakeout operations. The resin-sand systems used in molding and coremaking are the source of various gases and vapors reflecting the composition of the resin-catalyst system as shown in Table 2.6-4. (8) If sea coal is used in the molding sand, there may be a serious exposure to carbon monoxide. Few published data are available on the exposure of shakeout operators to toxic metal particulate resulting from sand contamination, although this must be considered a potential problem.

The principal control of air contaminants at the shakeout is local exhaust ventilation utilizing enclosures and side- or down-draft hoods (5). With the exception of well-designed enclosures for automatic shakeout of small castings, none of the systems are completely effective. Many approaches have been considered to replace or augment the conventional shakeout. Rather than vibrating the casting free from the molding sand, hydroblast units have been utilized to free the casting while generating little dust. Small molds may be placed in a special tumbler where the parts are separated from the sand. Failing good control, the shakeout is conducted during an off-shift to reduce the number of persons exposed.

The removal of cores from the castings is either accomplished at the shakeout or at a special core knockout station. High pressure water streams have been successfully applied in this operation.

2.6.5 Cleaning and Finishing

After shakeout the casting is processed in the finishing room. Complex cores require special core knockouts that may be either manual or mechanized operations depending on the core geometry and production rate. The extraneous metal including sprue, risers, and gates on gray iron castings are removed by a sharp rap of a hammer; burning or cutoff wheels are used on steel and alloys castings. The parts are frequently cleaned by abrasive blasting or hydroblast and rough finishing is done by chipping and grinding. If the castings are clean, there is limited exposure to silica sand during finishing; however, in steel foundries the sand will fuse on the surface and some of the silica may be converted to tridymite or cristobalite.

The occupational health problems in this area including dust exposure, noise, and vibration, are described in Sections 2.1 and 2.7. Silica dust is the major health hazard in the cleaning room.

2.6.6 Control Technology

Although other air contaminants exist, silica and carbon monoxide continue to be the major health hazards in the foundry. Noise is the most important physical stress.

As shown in Table 2.6-1, the hazard from silica varies with the physical state of the sand, its chemical composition, and the method of handling. Since the sand is initially dry, the mixing or mulling operation is dusty, and local exhaust ventilation is required even though the sand is coarse. After this point, the moisture content is high and little dusting occurs during the molding operation. After casting, the sand becomes friable and dusty; serious dust exposures may occur during shakeout and sand conditioning.

Historically, the approach to dust control in sand foundries has been the use of control ventilation coupled with housekeeping and wet methods (9). Low volume–high velocity capture systems are finding increased application on grinding and chipping operations in finishing rooms. In recent years the use of nonsiliceous parting compounds instead of silica flour has demonstrated the value of substitution as a control. Changes in procedure such as the Schumacher method, which mixes some prepared moist molding sand with dry shakeout sand before it is conveyed back from the shakeout, result in impressive dust reduction. The introduction of permanent mold techniques will, of course, have a major impact on silica sand usage and resulting worker exposure.

Table 2.6-5 *Noise Exposures in Foundry*

Operation	Sound Level (dBA)		Permissible Exposure Time (hr)[a]	
	Min.	Max.	Min.	Max.
Shakeout	105	115	1	0.25
Quick release–high pressure air	95	117	4	0.25
Air blowoff	100	120	2	0.25
Tumblers, unlined	100	115	2	0.25
Shotblast booths, outside booths	100	110	2	0.5
Chipping and grinding lines	95	115	4	0.25
Hopper vibrators	95	115	4	0.25
Molding machines	95	115	8	0.25
Millroom, general area	95	110	4	0.50
Man cooler fans	90	95	8	4
Sand slingers	90	95	8	4
Sand grinders	95	110	4	0.5
Wheelabrators®, loading, dumping	95	115	4	0.25
Hand ramming with air hammer	92	97	6	3

Source: Reference 9.
[a]Based on the OSHA standard of 90 dBA for an 8 hr day.

Foundry noise is a major problem, as shown in Table 2.6-5. The American Foundrymen's Society has published an excellent review of current knowledge of noise control on chippers and grinders, burners, electric arc furnaces, shakeout, molding operations, core and molding machines, and conveyors. This document treats the most difficult problems in the foundry and provides state-of-the-art technology solutions that have been evaluated in the industry (10).

Limited attention has been given the occurrence of vibration-induced disease such as Vibration-Induced White Finger (VWF) in the foundry population. A guide published in the United Kingdom recommends exposure limits for hand transmitted vibration from vibratory tools of the type used in the foundry (11).

REFERENCES

1 "Systems Analysis of Emission and Emission Control In The Iron Foundry Industry. Vol. II. Exhibits", U.S. Environmental Protection Agency Publication No. AP _____, EPA, Research Triangle, NC, Feb. 1971.

*2 American Foundrymen's Society, *AFS Foundry Environmental Control*, Vol. 1, AFS, Des Plaines, IL, 1972.

3 *Foundry Health and Safety Guide Series*, American Foundrymen's Society, Des Plains, IL, 1976.

4 "Health Risks in the Foundry," *Mich. Occup. Health,* **13**, 4, (Winter 1968).

5 J. H. Hagopian and E. K. Bastress, "Recommended Industrial Ventilation Guidelines," Final Report DHEW, Contract CEC-99-74-33, NIOSH, Cincinnati, OH, 1976.

6 Committee on Industrial Ventilation, American Conference of Governmental Industrial Hygienists, *Industrial Ventilation: A Manual of Recommended Practice,* 16th ed., ACGIH, Lansing, MI, 1980.

*7 "An Evaluation of Occupational Health Hazard Control Technology For the Foundry Industry," Department of Health, Education and Welfare, Publication No. (NIOSH) 78-114, Cincinnati, OH, 1978.

8 W. D. Scott, C. E. Bates, and R. J. James, *AFS Transactions* **85**, 203 (1977).

9 "How to Keep a Foundry Clean, Part 1," *Mich. Occup. Health,* **9**, 2 (Winter 1963–1964).

10 *State-of-the-Art Noise Control For Foundry Operations,* 2nd ed., American Foundrymen's Society, Des Plaines, IL, 1979.

11 "Guide to the Evaluation of Exposure of the Human Hand Arm System to Vibration," DD 43, British Standards Institution, London, 1975.

2.7 GRINDING, POLISHING, BUFFING

These operations are grouped together for discussion because they all involve controlled use of bonded abrasives for metal finishing operations and in many cases the three operations are conducted in the sequence noted. This discussion will cover the nonprecision applications of these techniques shown in Table 2.7-1 (1).

2.7.1 Processes and Materials

Nondimensional application of grinding techniques includes cutoff operations in foundries where gates, sprues, and risers are removed, rough grinding of forgings and castings, facing off weldments, and grinding out major surface imperfections in metal fabrications. Grinding is done with wheels of various geometries made up of selected abrasives in bonding structural matrices (1). The common abrasives are aluminum oxide and silicon carbide. A variety of bonding materials are available to provide mechanical strength and yet release the spent abrasive granules to renew the cutting surface. Vitrified glass is a common bonding agent available in a range of strenth and hardness. Sodium silicate (water glass) is the softest bonding agent, and for this reason it is used on grinding wheels for hard metals that require high wheel wear. Shellac and rubber bonding agents are routinely used for thin cutoff wheels that require maximum flexibility. Resinoid wheel bonds, based on thermosetting resins such as phenol-formaldehyde, may be reinforced with metal or fiberglass for heavy duty applications. In the past, lead was used in the wheel composition to improve lubricity and cooling. In general this practice has been curtailed, although an occasional manufacturer may continue to supply such products. The abrasive industry uses the standard labeling nomenclature shown in Figure 2.7-1 to identify the grinding wheel design (2). This information is valuable in determining the possible air contaminants released from the grinding wheel.

Table 2.7-1 *Source of Airborne Contaminants from Grinding, Polishing, and Buffing*

Equipment Types	Equipment Subtypes
Surface-type grinders	Surface grinders Roll grinders Snaggers Slab and billet grinders Swing grinders
Pedestal-type grinders	Pedestal grinders Bench grinders Floorstand grinders Tool grinders
Disc grinders and polishers	Single spindle disc grinders Double spindle disc grinders Disc polishers
Internal grinders	Internal grinders
Abrasive cutting-off machines	Abrasive cutting-off machines
Pedestal-type polishers and buffers	Wheel and drum polishers Backstand idler polishers Buffing lathes
Belt polishers	Belt grinders and polishers (using flat belt surface)
Portable grinders, polishers, and buffers	Portable grinders Portable polishers Portable buffers
Multiple polishers and buffers	Multiple-belt polishers Multiple-head buffers

Source: Reference 1.

Little information is available on the generation rate of grinding wheel debris on various applications; however the wheel components normally make up a small fraction of the total airborne particulates released during grinding. The bulk of the particulates are released from the workpiece. After use, the grinding wheel may load or plug and the wheel must be "dressed" with a diamond tool or "crushed dressed" with a steel roller. During this brief period, a significant amount of the wheel is removed and a small quantity may become airborne.

Polishing techniques are used to remove workpiece surface imperfections such as tool marks. This technique may be used to remove as much as 0.1 mm of stock from the workpiece. The abrasive, again usually aluminum oxide or silicon carbide, is bonded to the surface of a belt, disk, or wheel structure in a

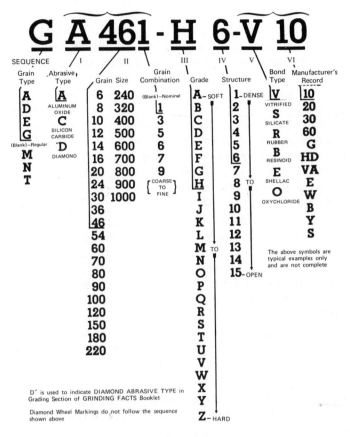

Figure 2.7-1 Identification of grinding wheels.

closely governed geometry and the workpiece is commonly applied to the moving abrasive carrier by hand.

The buffing process differs from grinding and polishing in that little metal is removed from the workpiece. The process merely provides a high luster surface by smearing any surface roughness with a lightweight abrasive. Red rouge (ferric oxide), and green rouge (chromium oxide) are used for soft metals; aluminum oxide, for harder metals. The abrasive is blended in a grease or wax carrier that is packaged in bar or tube form. The buffing wheel is made from cotton or wool discs sewn together to form a wheel or "buff." The abrasive is applied to the perimeter of the wheel and the workpiece is then pressed against the rotating wheel. The wheels are normally mounted on a buffing lathe that is similar to a grinding stand. During heavy duty operations, the surface temperature of the workpiece may reach 150°C (300°F).

2.7.2   Exposures and Control

The hazard potential from grinding, polishing, and buffing operations depends on the specific operation, the workpiece metal and its surface coating, and the type of abrasive system in use. A NIOSH-sponsored study of the ventilation requirements for grinding, polishing, and buffing operation shows that the major source of airborne particulates in grinding and polishing is the workpiece, while the abrasive and its support represent the principal source of contamination in buffing (2). Few data are available on the influence of the metal type on airborne levels of contaminants although limited tests on three metals in the aforementioned study showed that aluminum generated the least dust, a 1018 steel alloy was the next dustiest, and titanium the most dusty. Certainly the type of metal worked and the construction of the abrasive system govern the generation rate and the size characteristics of the dust; however, the data necessary for a detailed assessment of this matter are not available.

A listing of the alloys worked and information on the nature of the materials released from the abrasive system is needed in order to evaluate the exposure of the operator. In many cases the exposure to dust can be evaluated by means of personal air sampling with gravimetric analysis. If dusts of toxic metals are released, then specific analyses for these contaminants are necessary.

A major question arises as to the need for local exhaust ventilation on grinding operations. British authorities (3) state that control is required if one is grinding:

1   Toxic metals and alloys of such metals as antimony, beryllium, chromium, cobalt, lead, nickel, tungsten, and vanadium. + uranium

2   Ferrous and nonferrous castings produced by sand molding. The fused sand on such castings is released during grinding with resulting exposure to silica.

3   Metal surfaces coated with toxic material, such as lead- and chromate-based paints.

4   Metals giving rise to substantial quantities of dust and fume.

The same source also states that control is not needed if grinding steel surfaces before welding, forging, stamping, and precision grinding. This statement may apply to mild steel but if alloy steels are worked, ventilation may be required on these operations. Control of exposures from these operations should include the removal of toxic coatings from the metal workpiece by such techniques as abrasive blasting under controlled conditions.

One can extend the application of the preceding guidelines on grinding to polishing. The ventilation requirements for buffing, however, are principally based on the large amount of debris released from the wheel, which may present a housekeeping problem and a potential fire risk.

The principal dust control technique on grinding, polishing and buffing operations is ventilation. The hood designs are based on a tight hood enclosure with minimum wheel-hood clearance to provide dust control at a minimum exhaust volume. An adjustable tongue on the hood also reduces the air induced by wheel rotation at high speed. On buffing operations, the hood is usually designed with a settling chamber to minimize plugging of the duct with wheel debris.

The hazard from bursting wheels operated at high speeds and the fire hazard from handling certain metals such as magnesium affect the design of hoods and wet dust collection systems.

The conventional ventilation control techniques for grinding, polishing, and buffing are well described in the ACGIH and ANSI publications (4, 5). A detailed approach by Hagopian and Bastress has been adopted in a NIOSH study of ventilation guidelines (6).

Fixed location operations are handled by conventional ventilation systems, whereas portable tools can be efficiently exhausted by low volume-high velocity systems. These techniques are widely used in metal finishing in the United Kingdom (7) and have seen limited use in beryllium machining and asbestos fabrication in the United States (8). The author's laboratory evaluated the effectiveness of a low volume-high velocity system mounted on a disc sander used on a rubber sonar dome containing a tributyl tin oxide fungicide, and found it effective in minimizing exposure to this toxic organic tin compound. The low air flow rates required by this technique will result in its increasing application for both fixed and portable installation as we move toward energy conservation.

The performance of the conventional hoods on grinding, polishing, and buffing operations should be checked periodically. Since the hood entry loss factor for conventional hoods is known, the hood static suction method is a quick, efficient way to evaluate the exhaust rates (4).

As in other cases, the use of water on grinding operations may not eliminate the need for ventilation control. Water droplets can evaporate, leaving a respirable airborne particle.

REFERENCES

1 E. Bastress, J. Niedzwecki, and A. Nugent. "Ventilation Requirements for Grinding, Buffing, and Polishing Operations" Department of Health, Education, and Welfare, Publication No. (NIOSH) 75-107, Cincinnati, OH, 1975.

2 American National Standards Institute, "Markings for Identifying Grinding Wheels and Other Bonded Abrasives," B.74.18-1977, ANSI, New York, 1977.

3 "Control of Dust From Portable Power Operated Grinding Machines (Code of Practice)," United Kingdom Department of Employment, London, 1974.

4 Committee on Industrial Ventilation, American Conference of Governmental Industrial Hygienists, *"Industrial Ventilation: A Manual of Recommended Practice,"* 16th ed., ACGIH, Lansing, MI, 1980.

5 American National Standards Institute, "Ventilation Control of Grinding, Polishing, and Buffing Operations," Z43.1-1966, ANSI, New York, 1966.

*6 J. H. Hagopian and E. K. Bastress, "Recommended Industrial Ventilation Guidelines," NIOSH Research Contract Report No. CDC-99-74-33, Cincinnati, OH, 1974.

7 "Dust Control: The Low Volume-High Velocity System," Technical Data Note 1, 2nd rev., Department of Employment, HM Factory Inspectorate, London, 1974.

8 W. A. Burgess and J. Murrow, *Am. Ind. Hyg. Assoc. J.*, **37**, 546 (1976).

2.8   HEAT TREATING

A range of heat treating methods for metal alloys is available to improve the strength, impact resistance, hardness, durability, and heat and corrosion resistance of the workpiece (1). In the most common procedures, metals are hardened by heating the workpiece to a high temperature with subsequent rapid cooling. Softening processes normally involve only heating, or heating with slow cooling.

2.8.1   Hardening

Case hardening, the production of a hard surface or case to the workpiece, is normally accomplished by diffusing carbon or nitrogen into the metal surface to a given depth to achieve the hardening of the alloy. This process may be accomplished in air, in atmosphere furnaces, or in immersion baths by one of the following methods.

Carburizing

In this process the workpiece is heated to 870–980°C (1600–1800°F) in an atmosphere containing high concentrations of carbon monoxide which is the source of the diffused carbon. In gas carburizing, the parts are heated in a furnace containing hydrocarbon gases or carbon monoxide while in pack carburizing the part is covered with carbonaceous material that burns to produce the carbon monoxide gas blanket. A third technique, retort carburizing, is identical to pack carburizing except that the carbonaceous fuel is not in direct contact with the part.

The heat treating furnaces used in gas carburizing are classified by the American Gas Association as shown in Table 2.8-1 (1). The furnace atmospheres normally used in gas carburizing are Classes 100, 300, 400, and 500. Carbon monoxide must be regulated closely to achieve concentrations that permit carbon to diffuse, but that do not scale or disfigure the steel.

The principal occupational health hazard in carburizing is the exposure to the products of combustion, especially carbon monoxide. In gas carburizing the furnace atmosphere is supplied by an atmosphere generator. In its simplest form, the generator burns or reacts the gas fuel under controlled conditions to

Table 2.8-1 *Heat Treating Furnace Atmospheres*

Class No.	Generation Technique	Typical Composition (%)
100	Combustion of hydrocarbons in air	70–80%N_2 balance CO, CO_2, H_2
200	Remove CO_2 and water from Class 100	97%N_2
300	Partial reaction of fuel gas and air with heat and catalyst to crack hydrocarbons	40%N_2 20%CO 40%H_2
400	Air passed over incandescent charcoal	65%N_2 35%CO
500	Burn fuel gas and air, remove water vapor, convert CO_2 to CO	—
600	Raw NH_3, dissociated NH_3, or burned NH_3	75%H_2 25–99%N_2

Source. Reference 1.

produce the correct concentration of carbon monoxide, and this gas is supplied to the furnace. Since carbon monoxide concentrations up to 40% may be used, small leaks may result in significant workroom exposure. Workplace exposures to carbon monoxide greater than 100 ppm are not uncommon in carburizing installations. In addition, if the parts are not clean, volatile material will be driven off in the furnace and generate toxic air contaminants that may leak into the workplace.

To control emission from carburizing operations the various combustion processes must be closely controlled, furnaces should be maintained in tight condition, dilution ventilation installed to remove fugitive leaks, furnaces provided with flame curtains at doors to control escaping gases, and self-contained breathing apparatus should be available for escape and repair operations (2).

Cyaniding

The conventional method of liquid carbonitriding is immersion in a cyanide bath with a subsequent quench. The part is commonly held in a sodium cyanide bath at temperatures above 870°C (1600°F) for 30–60 min. The air contaminant released from this process is sodium carbonate; cyanide compounds apparently are not released, although there are no citations in the open literature demonstrating this. Local or dilution ventilation is frequently applied to this process, although standards have not been proposed. The handling of cyanide salts requires the same precautions as those noted in Section 2.4, that is, secure and dry storage, isolation from acids, and planned disposal of waste. Care must be taken in handling quench liquids since the cyanide salt residue on the part will, in time, contaminate the quench liquid.

Gas nitriding

Gas nitriding is a common means of achieving hardening by the diffusion of nitrogen into the metal. This process utilizes a furnace atmosphere of ammonia operating at 510–570°C (950–1050°F) noted as Class 600 in Table 2.8-1. The handling of ammonia in this operation is hazardous in terms of fire, explosion, and toxicity.

2.8.2 Annealing (Tempering Baths)

These salt baths may be nitrate salts alone or based on an equal mix of nitrate and nitrite salts operating in the range of 200–480°C (400–900°F) (3). The principal application for these low temperature baths is for aluminum or special working of steel alloys. Rigorous storage and handling precautions are necessary for the sodium and potassium nitrate salts since they are powerful oxidizing agents. Nitrate salts may start to decompose at 400°C (750°F); at 650°C (1200°F) the decomposition may be violent with the release of oxides of nitrogen to the workplace. The bath container itself may be destroyed under such conditions, releasing the contents to the workplace and presenting a major chemical spill problem and a fire and explosion hazard if the hot nitrate salts contact organic materials such as carbon or grease.

The bath temperature controls must be reliable and the bath must be equipped with an automatic shutdown. Before the bath is brought down to room temperature, rods should be inserted into the metal bath. These rods will provide vent holes to release gases when the bath is again brought up to temperature. If this is not done and gas is occluded in the bath, explosions or blowouts may occur. All parts must be clean and dry before immersion in the nitrate bath since residual grease, paint, and oil may cause explosions. Alloys containing more than a few percent magnesium should not be processed in these baths. Nitrate salt storage should be in a locked, secure area that is dry and free from organic material. These baths may require local exhaust ventilation although no design data are available.

2.8.3 Quenching

As mentioned in Section 2.8.1, hardening of metal alloys requires a rapid quench after the high temperature bath. The quench baths may be water, oil, molten salt, liquid air, or brine. The potential problems range from a nuisance problem due to release of steam from a water bath to acrolein or other thermal degradation products from oil (4). Local exhaust ventilation may be necessary on oil quench tanks.

2.8.4 Hazard Potential

The principal problems in heat treating operations are due to the special furnace environments, especially carbon monoxide, and the special hazards from handling nitrate and cyanide bath materials. Although the hazard potential is significant from these operations, few data are available. In the only published data on air contaminants in metal annealing and hardening operations, Elkins noted hazardous concentrations for carbon monoxide in 19 out of 108 installations, and 59 out of 167 exceeded the then current maximum allowable concentration for lead (5). Hazardous cyanide concentrations were not observed.

Lead baths are frequently held at temperatures between 540°C (1000°F) and 820°C (1500°F), therefore requiring exhaust ventilation. Significant lead exposures may occur when removing the surface oxidation or dross floating on this molten lead. Oil quench tanks are often ventilated to remove the irritating smoke that evolves during their use. Some tanks have cooling coils to control the temperature of the oil and to reduce both smoke production and the fire hazard. There are no ventilation standards published for these operations.

Where sprinkler systems are used, canopies should be erected above all oil, salt, and metal baths to prevent water from cascading into them. Any workman who happened to be adjacent to a hot bath when water struck it would be in grave danger.

REFERENCES

*1 *Guide To Heat Treating Services,* 2nd ed., Metal Treating Institute, Tallahassee, FL, 1976.

2   J. Danielson, Ed., *Air Pollution Engineering Manual,* 2nd ed., Publication No. AP-40, Government Printing Office, Washington, D.C., 1973.

3   American Mutual Insurance Alliance, Specific Industrial Processes, "Heat Treating—Nitrate Salt Baths," Chicago, IL, 1951.

4   A. D. Brandt, *Industrial Health Engineering,* Wiley, New York, 1947.

5   H. B. Elkins, *The Chemistry of Industrial Toxicology,* Wiley, New York, 1959.

2.9 INDUSTRIAL RADIOGRAPHY (1, 2, 3)

Radiography is used principally in industry for the examination of metal fabrications such as weldments, castings, and forgings in a variety of settings. Specially designed shielded cabinets may be located in manufacturing areas for in-process examination of parts. Large components may be transported to shielded rooms for examination. Radiography may be performed in open shop areas, on construction sites, on board ships, and along pipelines.

The process of radiography consists of exposing the object to be examined to x rays or gamma rays from one side and measuring the amount of radiation that emerges from the opposite side. This measurement is usually made with film or a fluoroscopic screen to provide a visual, two dimensional display of the radiation distribution.

The process is used to locate inconsistencies in the object or part. As radiation passes through the part, some of the radiation is absorbed or attenuated. The degree of attenuation depends on the thickness of the part, its density, and its atomic number. For example, metal castings frequently contain subsurface porosities. These can be found by radiography because the radiation passing through the section of the casting containing the porosity will be attenuated to a lesser degree than the radiation passing through a solid section of the casting, as shown in Figure 2.9-1. This difference in attenuation of the radiation results in differences in optical density in the processed film.

The principal potential hazard in industrial radiography is exposure to ionizing radiation. This section deals with safety precautions designed to minimize worker exposure to radiation. Since many of the specific safety considerations relate to the type of radiation source (x rays or gamma rays), these sources will be treated independently.

2.9.1 X-ray Sources

X rays used in industrial radiography are produced electrically and therefore fall into the category of "electronic product radiation." The most commonly used source of x rays is the conventional x-ray generator. This device consists of an evacuated tube in which electrons are accelerated through a high potential difference from a cathode to an anode. The anode contains a target fabricated from a material of relatively high atomic number (usually tungsten). As electrons impinge on the target, they rapidly decelerate, producing

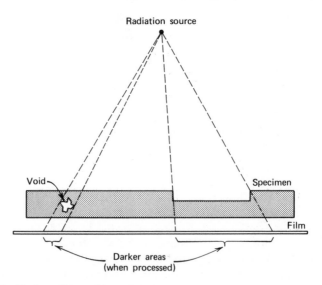

Figure 2.9-1 Basic radiographic process.

bremstrahlung radiation (x rays). Conventional x-ray generators used in industrial radiography produce peak x-ray energies from 40 to 420 keV.

The x-ray tube is housed in a tube head or x-ray head as shown in Figure 2.9-2. This head also contains high voltage transformers and insulating material. Many industrial x-ray tube heads have integral lead shielding to reduce the radiation intensities in directions other than the direction of interest. Table 2.9-1 lists typical radiation intensities produced by conventional industrial x-ray generators. This table also lists typical sizes and weights for these tube heads. The control panel contains x-ray energize and deenergize switches, usually an "x-ray on" indicator, panel meters that indicate the tube potential and tube current, and means for adjusting these parameters. The control panel and interconnecting cable to the tube head allow the operator to energize the x-ray tube from a remote location in order to minimize radiation exposure.

Higher energy x rays are generated by Van de Graaf accelerators (up to a few MeV), linear accelerators (up to approximately 10 MeV), and betatrons (up to 25 MeV). In these devices, x rays are also produced by decelerating electrons. However, the method of accelerating these electrons differs among devices. Van de Graaf accelerators are used fairly widely in industrial radiography, although they are much less commonly used than conventional x-ray generators.

Figure 2.9-2 X-ray head.

Table 2.9-1   *Characteristics of Typical Industrial X-ray Generators*

Tube Potential (kV)	Length (mm)	Diameter (mm)	Mass (kg)	Dose Rate at 1m from Focal Point (R/hr)
160	580	188	30	560
200	651	221	47	900
250	900	243	78	1300
300	1000	263	103	1400

Because of the size, weight, and services required, it is difficult to position conventional x-ray generators and accelerators to perform radiography in field locations. Consequently, these devices are used most widely in fixed installations such as shielded exposure rooms.

The design and manufacture of industrial x-ray generators are regulated by the Food and Drug Administration, Bureau of Radiological Health. The American National Standards Institute has developed a standard for the design and manufacture of these devices (4). These standards specify maximum allowable radiation intensities outside the useful beam. They require warning lights on both the control panel and the tube head to indicate when x rays are being generated.

The use of industrial x-ray generators is regulated by OSHA. Additionally, many states regulate the use of industrial x-ray generators within their own jurisdictions. Areas in which radiography is performed must be posted with signs that bear the radiation caution symbol and a warning statement (Figure 2.9-3). Access to these areas must be secured against unauthorized entry. Occasionally, radiation exposures occur to nonradiographic personnel because they unknowingly enter radiography areas. When radiography is being performed in open manufacturing areas, it is essential to instruct other workers in the identification and meaning of these warning signs in order to minimize these occurrences.

Radiographic operators are required to wear personnel monitoring devices to measure the magnitude of their exposure to radiation. Typical devices are film badges, thermoluminescent dosimeters, and direct reading pocket dosimeters. It is also advisable for radiographic operators to use audible alarm dosimeters or "chirpers." These devices emit an audible signal or "chirp" when exposed to radiation. The frequency of the signal is proportional to the radiation intensity. They are useful in warning operators who unknowingly enter a field of radiation.

Figure 2.9-3 Posting of area for radiography.

The operators must be trained in use of radiation survey instruments to monitor radiation levels to which they are exposed and to assure that the x-ray source is off at the conclusion of the operation. A wide variety of radiation survey instruments is available for use in industrial radiography. When using industrial x-ray generators, especially relatively low energy generators, it is imperative that the instrument has an appropriate energy range for measuring the energy of radiation used.

Radiation exposure is not the only potential hazard associated with industrial x-ray generators. These devices use high voltage power supplies for the production of x rays, and consideration must be given to the electrical hazards associated with using this equipment.

2.9.2 Gamma Ray Sources

Gamma rays used in industrial radiography are produced as a result of the decay of radioactive nuclei. The principal radioisotopes used in industrial radiography are iridium-192 and cobalt-60. Other radioisotopes, such as ytterbium-169 and thulium-170, are much less common but have some limited applicability. Radioisotope sources produce gamma rays with discrete energies as opposed to the continuous spectrum of energies produced by x-ray generators. Due to radioactive decay, the activity of radioisotope sources decreases exponentially with time. Consequently, the radiation intensities from these sources similarly decrease. Table 2.9-2 lists the typical physical parameters of these sources. Typical radiographic sources contain up to 200 Ci of iridium-192 and up to 1000 Ci of cobalt-60.

Table 2.9-2 *Characteristics of Typical Radiographic Radioisotopes*

Isotope	Half-Life	Principal Photon Energies (keV)			Dose Rate at 1 m from 1 Ci (R/hr)
^{192}Ir	74 days	310	468	608	0.55
^{60}Co	5.3 years	1170	1330		1.3
^{169}Yb	30 days	60	110	131	0.125
		177	198		
^{137}Cs	30 years	660			0.32

Unlike x-ray generators, radioisotope sources require no external source of energy, whch makes their use attractive in performing radiography in remote locations, such as on pipelines. Since they are not energized by an external power supply, these sources cannot be "turned off," and they continuously emit gamma rays. For this reason, certain additional safety precautions must be exercised.

In industrial radiographic sources, the radioisotope is sealed inside a source capsule that is usually fabricated from stainless steel. The radioisotope source capsule is stored inside a shielded container or "pig" when not in use to reduce the radiation intensities in the surrounding areas. These containers weigh up to 50 lb for housing iridium-192 sources and up to several hundred pounds for cobalt-60 sources. A flexible tube is attached to the storage container with a source stop at the end of the tube (Figure 2.9-4). The source stop is located at the irradiation position by the operator, and the radioisotope source is transferred through the flexible tube to the source stop by means of remote handling equipment. This remote handling equipment allows the operator to be as far as 17 m (50 ft) from the source in order to minimize his or her radiation exposure. At the conclusion of the radiographic exposure, the operator remotely retracts the source from the source stop to the shielded position in the exposure device. As already noted, radioisotope sources are generally favored for use in field construction sites or cramped environments such as shipboard because positioning the lightweight source stop is more convenient than positioning an x-ray generator.

The design, manufacture, and use of radioisotope sources and exposure devices for industrial radiography are regulated by the U.S. Nuclear Regulatory Commission. However, the Nuclear Regulatory Commission has entered into an agreement with 26 states that allows these "Agreement States"

Figure 2.9-4   Radioisotope storage unit.

to regulate radioisotope radiography within their jurisdictions. Organizations that wish to perform radioisotope radiography must obtain a license from either the Nuclear Regulatory Commission or the Agreement State. In order to obtain this license, they must describe their safety procedures and equipment to the licensing authority.

Radiographic operators must receive training as required by the regulatory bodies including instruction in their own organization's safety procedures, a formal radiation safety training course, and a period of on-the-job training where the individual works under the direct personal supervision of a qualified radiographer. At the conclusion of this training, the operator must demonstrate his knowledge and competence to the licensee's management.

Areas in which radiography is being performed must be posted with radiation warning signs, and access to these areas must be secured as described earlier. An operator performing radioisotope radiography must wear both a direct reading pocket dosimeter and either a film badge or a thermoluminescent dosimeter. Additionally, the operator must use a calibrated radiation survey instrument during all radiographic operations. In order to reduce the radiation intensity in the area after an exposure, the operator must retract the source to a shielded position within the exposure device. The only method available to the operator for assuring that the source is shielded properly is a radiation survey of the area. The operator should survey the entire perimeter of the exposure device and the entire length of guide tube and source stop after each radiographic operation to assure that the source has been fully and properly shielded.

Occasional radiation exposure incidents have occurred in industrial radioisotope radiography. These incidents generally result because the operator fails to properly return the source to a shielded position in the exposure device and then approaches the exposure device or source stop without making a proper radiation survey. The importance of making a proper radiation survey at the conclusion of each radiographic exposure cannot be overemphasized.

The frequency of radiation exposure incidents may be reduced through the use of audible alarm dosimeters. As described earlier, these devices emit an audible signal when exposed to radiation. In cases where operators have not properly shielded sources and fail to make a proper survey, the audible alarm dosimeter might alert the operator to abnormal radiation levels. Some current types of radiographic exposure devices incorporate source position indicators that provide a visual signal if the source is not stored properly. Use of these devices may also reduce the frequency of radiation exposure incidents.

REFERENCES

1 NUREG—0495 Public Meeting on Radiation Safety for Industrial Radiographers, U.S. Nuclear Regulatory Commission, Washington, D.C., 1978.

2 *Isotope Radiography Radiation Safety Handbook,* Technical Operations, Inc., Burlington, MA, 1981.

3 "A Radiological Health Study of Industrial Gamma Radiography in Canada," Department of National Health and Welfare, Ottawa, Canada, 1979.

*4 American National Standards Institute, "Radiological Safety Standard for the Design of Radiographic and Fluoroscopic Industrial X-ray Equipment," ANSI/NBS 123-1976, ANSI, New York, 1976.

2.10 METAL MACHINING

The fabrication of metal parts from solid stock is done with a variety of machine tools the most common of which are the lathe, drill press, miller, shaper, planer, and surface grinder. The occupational health hazards from these operations are similar, so they will be grouped together under conventional machining. Two of the less conventional machining techniques, electrochemical and electrical discharge machining, are also discussed in this section.

2.10.1 Conventional Tool Machining

The major machining operations of turning, milling, and drilling utilize cutting tools that shear off the metal from the workpiece forming a thin running coil

- Cutting Fluids (dermatitis, sensitization)
· Oil mist (nitrosamines)
· Chips Insuring eyes-skin

that normally breaks at the tool to form small chips. Extremes of temperature and pressure occur at the interface between the cutting tool and the work. To cool this point, provide an interface lubricant, and help flush away the chips formed by the cutting tool, a coolant or cutting fluid is directed to the cutting point in a solid stream (flood) or a mist (1).

The airborne particulates generated by these machining operations depend on the type of base metal and cutting tool, the dust forming characteristics of the metal, the machining technique, and the coolant and the manner in which it is applied. Each of these concerns will be addressed briefly in this section.

The type of metal being machined is, of course, of paramount concern. The metals range from mild steel with no potential health hazard as a result of conventional machining to various high temperature and stainless alloys incorporating known toxic metals including chromium, nickel, and cobalt that may present low airborne exposures depending on the machining technique. Finally, highly toxic metals such as beryllium do present significant exposures that require vigorous control in any machining operation as noted in Chapter 3. Under normal machining operations using cutting tools, the metal particulate airborne concentration from conventional metals and alloys is minimal, although the particulate generation rates and the airborne particle sizes from various metals for given machining operations have not been documented.

A range of specialized alloys has been developed for use in the manufacture of cutting tools (1, 2). These materials include: (a) high carbon steels with alloying elements of vanadium, chromium, and manganese; (b) high speed steels containing manganese and tungsten; (c) special cobalt steels; (d) cast alloys of tungsten, chromium, and cobalt; and (e) tungsten carbide. The loss of material from the cutting tool is insignificant during conventional machining and does not represent a potential hazard. Preparing the cutting tools for application may involve a significant exposure to toxic metal dusts during grinding and dressing operations and such operations should be provided with local exhaust ventilation (3).

Coolants and cutting fluids, important components in metal machining, are designed to cool and lubricate the point of the cutting tool and flush away chips (4, 5). These fluids are presently available in the form of: (a) soluble (emulsified) cutting oils based on mineral oil emulsified in water with soaps or sulfonates; (b) straight cutting oils based on mineral oils with the addition of fatty acids; and (c) synthetic oils of varying composition. A broad description of the composition of the three types of cutting fluids is shown in Table 2.10-1; a list of various additives for special functions is noted in Table 2.10-2; and a general review of the application realm for these fluids is outlined in Table 2.10-3.

Cutting fluids present two potential health problems: extensive skin contact with the cutting fluids and the inhalation of the respirable oil mist. It has been estimated that over 400,000 cases of dermatitis occur in the United States each year as a result of contact with coolants and cutting fluids. Oils may cause acne lesions, infection of sweat pores, and hair folliculitis. One also notes occasional sensitization to these materials. Petroleum oils may be direct irritants if the boiling point is less than 350°C (660°F), while those with boiling points above 350°C normally cause folliculitis and oil acne (6, 7). Key has stated that the bacterial contamination of the fluids is not a major cause of dermatitis, although this allegation continues to be made (6). Bacterium isolated from cutting fluid sumps have not been identified as hazardous to man.

Table 2.10-1 *Cutting Fluid Composition*

Mineral Oil
Base 60–100%, paraffinic or naphthenic
Polar additives
  Animal and vegetable oils, fats, and waxes to wet and penetrate the chip/tool interface
  Synthetic boundary lubricants: esters, fatty oils and acids, poly or complex alcohols
Extreme pressure lubricants
  Sulfur-free or combined as sulfurized mineral oil or sulfurized fat
  Chlorine—as long chain chlorinated wax or chlorinated ester
  Combination—sulfo-chlorinated mineral oil or sulfo-chlorinated fatty oil
  Phosphorous—as organic phosphate or metallic phosphate
Germicides

Emulsified Oil (Soluble Oil)—Opaque, Milky Appearance
Base—mineral oil, comprising 50–90% of the concentrate; in use the concentrate is diluted with water in ratios of 1–5 to 1–50
Emulsifiers: petroleum sulfonates, amine soaps, rosin soaps, naphthenic acids
Polar additives—sperm oil, lard oil, and esters
Extreme pressure lubricants
Corrosion inhibitors: polar organics, for example, hydroxyl amines
Germicides
Dyes

Synthetics (Transparent)
Base—water, comprising 50–80% of the concentrate; in use the concentrate is diluted with water in ratios of 1–10 to 1–200. True synthetics contain no oil. Semisynthetics that contain mineral oil present in amounts of 5–25% of the concentrate are available.
Corrosion inhibitors
  Inorganics—borates, nitrites, nitrates, phosphates
  Organics—amines, nitrites
Surfactants
Lubricants—esters
Dyes
Germicides

Source: Reference 8.

Table 2.10-2 *Additives to Lubricants,*
Coolants, and Cutting Fluids

Extreme pressure agents
   Tricresyl phosphate
   Zinc dithiophosphate
   Chlorinated paraffins
   Chlorinated diphenyls
Antioxidant and anticorrosion inhibitors
   Zinc alkyl
   Aryl dithiophosphate
   Methyl ditertiary butyl phenol
   p-Phenylenediamine
   Chrome and sodium nitrite
Detergents and dispersants
   Methyl alkyl sulphonates
   Alkyl phenolates
Bactericides
   Phenolic compounds
   Quaternary ammonium compounds

Source.   Reference 1.

Recent studies have shown that the fluids may become contaminated with trace quantities of the base metal that is being worked (9). These dissolved metals in the fluid are then contacted by the worker. It is possible that this phenomena occurs with alloys containing chromium, nickel, and cobalt, all strong sensitizers.

The occupational health literature from the United Kingdom stresses the potential skin cancer problem from contact with lubricating and mineral oils during machining operations. Although strong measures have been taken in the United Kingdom to combat this issue, including the introduction of solvent refined oils, the problem still exists in certain areas.

The application of the cutting fluid to hot, rotating parts causes an oil mist to be generated, which creates the characteristic smell in the machine shop environment. The introduction of mist application of fluids in the 1950s has contributed to this problem. Several investigators have demonstrated minimal medical effects from exposure to mists of straight cutting oils while others suggest it is harmful. Waldon presents a strong case for rigorous control of oil mist noting that a study population with scrotal cancer attributed to oil contact also had excess cancer of respiratory and digestive systems (10).

An excellent pamphlet on lubricating and coolant oils prepared by Esso (7) outlines a recommended procedure to minimize exposure to cutting fluids. The following selected material is quoted:

1 Avoid all unnecessary contact with mineral or synthetic oils. Minimize contact by using splash guards, protective gloves, and protective aprons, etc. Use goggles or face visors when handling soluble oil concentrate. Have

Table 2.10-3 *Application of Cutting Fluids*

Process	Tool	Workpiece Material							
		Mg Alloys	Al Alloys	Cu Alloys	Steel	Stainless Steel, Ni Alloys	Cast Iron	Ti Alloys	
Turning	HSS	Dry MO CH	MO-EM CH FA-MO	MO-EM CL-EM FO-MO	EP-EM CH	CL-MO CL-EM	EP-EM EP-MO CH^s	EP-EM EP-CH	
	WC		CH FA-MO Dry	MO-EM FO-MO CH	Dry EP-EM EP-CH	Dry CL-MO CL-EM	Dry EP-EM CH	EP-MO EP-EM	
Milling	HSS			MO-EM FO-MO CH	EP-MO EP-EM	CL-MO CL-EM	EP-EM CH	EP-EM EP-CH	
	WC				Dry EP-MO EP-EM	Dry CL-MO CL-EM	Dry EP-EM CH	EP-MO EP-EM	
Drilling	HSS		MO-EM FO-EM		EP-CH EP-MO EP-EM	CL-MO	EP-EM EP-CH	EP-MO	

90

	Workpiece Material					
WC	EP-EM FA-MO		EP-CH EP-EM	CL-MO CL-EM	Dry EP-EM EP-CH	
Reaming HSS Broaching Tapping			EP-MO EP-EM	CL-MO CL-EM	Dry EP-EM EP-CH	
Grinding	MO CH	CH FO-MO	EP-CH	CL-EM CL-CH	EP-CH EP-EM	EP-CH EP-EM

Source. Reference 4. From *Introduction to Manufacturing Processes* by J. A. Schey. Copyright 1977. Used with permission of McGraw-Hill Book Co.

CL = chlorinated paraffin.

EM = emulsion; the listed lubricating ingredients are finely distributed in water.

EP = "extreme-pressure" compounds (containing S, Cl, and P).

FA = fatty acids and alcohols, for example, oleic acid, stearic acid, stearyl alcohol.

FO = fatty oils for example, plam oil and synthetic palm oil

MO = mineral oil (viscosity in parentheses, in units of centipoise at 40°C)

CH = water-based chemical solutions and surface-active compounds

protective clothing that becomes contaminated dry cleaned. The golden rule is: don't wear oil-soaked clothing and never put oily rags into pockets.

2   Encourage workers to wear clean work clothes, since oil-soaked clothing may hold the oil in contact with the skin longer than would otherwise occur. This applies particularly to underclothes, which should be changed frequently and washed thoroughly before re-use. Consider the provision of one locker for work clothes and a separate one for street clothes.

3   Consider the use of short-sleeved overalls rather than long-sleeved garments for workers handling metalworking fluids where friction on the skin from cuffs saturated in oil and swarf can promote skin problems.

4   Assist removal of oil from the skin as soon as possible if contact does occur. This usually means the installation of easily accessible wash basins and the provisions of mild soap and clean towels in adequate supply. Avoid strong soaps and detergents and abrasive-type skin cleaners.

5   Encourage workers to take showers at the end of a day's work in order to remove all traces of oil from the skin.

6   Do not allow solvents to be used for cleansing the skin. Use only warm water, mild soap, and a soft brush or in combination with a mild proprietary skin cleanser.

7   Encourage the use of a skin reconditioning cream at the end of the shift, after washing hands. These products help to replace the natural fats and oils removed from the skin by exposure to oils and by washing, and are a very important part of a skin conservation programme.

8   Encourage the use of barrier cream before starting work, and also after each time hands are washed. Different barrier creams are needed to protect against different oils, so ensure that the correct one is used.

9   Avoid unncessary exposure of workers to oil mist or vapours. In any event, ensure that breathing zone levels of oil mist are well below the recommended permissible concentration of 5 milligrams/cubic metre air.

10   See that all cuts and scratches receive prompt medical attention.

11   Prevent contamination of all oils particularly the soluble oils, and minimize the use of biocides. Ensure that soluble oils are used only at the recommended dilution ratios.

12   Programme the regular cleaning of machines that use oil.

13   Obey any special instructions on product labels. In common with most other industries, the petroleum industry is increasing its use of precautionary labelling.

14   Use correct work technique—particularly for soluble oil concentrates which may irritate skin and eyes. Handling concentrate and preparing dilutions requires careful precautionary measures; the use of goggles or a face-visor, impervious gloves, etc. Use warning notices, placards, etc. to draw attention to the need for good personal hygiene and good work practices.

In many cases the machining operations on such metals as magnesium and titanium may generate explosive concentrations of dust, and the operations must be conducted with suitable ventilation control and air cleaning.

The exhaust hoods commonly used to capture toxic metal dusts from machining include conventional exterior hoods, high velocity–low volume capture hoods, and enclosures (11). Several special hood enclosures have been utilized in beryllium fabrication shops (see Section 3.9).

The discovery that cutting fluids containing both nitrates and amines could result in the generations of various carcinogenic nitrosamines prompted a NIOSH Intelligence Bulletin (12) and a field study of exposure to cutting fluid mists (8). Average concentrations of oil mist for a variety of machining operations ranged from 0.2 to 1.9 mg/m³. The authors stated that concentrations of oil mist can be maintained below the present TLV of 5 mg/m³ by the use of oil containing antimist additives, hood enclosures with suitable exhaust ventilation, and air cleaning. Since carcinogenic polyaromatic hydrocarbons were tentatively identified in certain bulk cutting oils, the authors recommended that appropriate chemical studies be conducted to evaluate this problem.

2.10.2 Electrochemical Machining (ECM)

The ECM process is almost a mirror image of the electroplating process described in Section 2.4. In electroplating, metal is deposited on the workpiece (cathode) from a solid piece of plating stock (anode) utilizing a dc electrolytic bath operating at low voltage and high current density. In the ECM process shown in Figure 2.10–1, the workpiece is the anode. The cutting tool, or

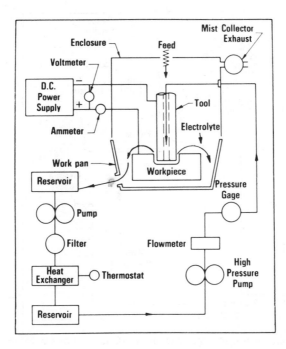

Figure 2.10-1 Electrochemical machining. (Courtesy of Chemform Division of KMS Industries.)

cathode, is machined to reflect the geometry of the hole to be cut in the workpiece. Electrolyte is pumped through the close space between the tool and the workpiece. As the dc current flows, metal ions are removed from the workpiece and are swept away by the electrolyte. Rather than depositing on the tool (cathode) the metal particles react with the electrolyte, usually an aqueous solution of sodium chloride or sodium nitrate, to form insoluble hydroxides that deposit out as a sludge. The tool is fed into the workpiece to complete the cut (13, 14).

Since the ECM method is fast, produces an excellent surface finish, does not produce burrs, and causes little tool wear, it is widely used for cutting irregularly shaped holes in hard, tough metals.

In the operation, the electrolyte is dissociated and hydrogen is released at the cathode. Local exhaust ventilation must be provided to ensure hydrogen concentrations do not approach the lower flammability limit. The equipment, which normally operates at low dc voltage and high current, does not present an electrical hazard.

2.10.3   Electrical Discharge Machining (EDM)

A spark-gap technique is the basis for this procedure, which is a popular machining technique for large precise work such as die sinking or the drilling of small holes in complex parts (14). In the system shown in Figure 2.10–2, a

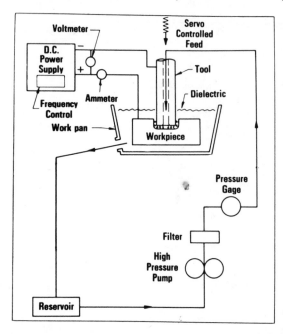

Figure 2.10-2   Electrical discharge machining. (Courtesy of Chemform Division of KMS Industries.)

graphite tool has been machined to the precise size and shape of the hole to be cut. The workpiece (anode) and the tool (cathode) are immersed in a dielectric oil bath and are powered by a low voltage dc power supply. The voltage across the gap increases until breakdown occurs and there is a spark discharge across the gap that produces a high temperature at the discharge point. This spark erodes a small quantity of the metal from the workpiece. The cycle is repeated at a frequency of 200 to 500 Hz with rather slow, accurate cutting of the workpiece.

The hazards from this process are minimal and are principally associated with the oil. In light cutting jobs, a petroleum distillate such as Stoddard Solvent is commonly used, while in large work a mineral oil is the dielectric. When heavy oil mists are encountered, local exhaust ventilation is needed. The oil gradually becomes contaminated with small hollow spheres of metal eroded from the part. As in the case of conventional machining, these metals may dissolve in the oil and present a dermatitis problem. An ultrahigh efficiency filter should be placed in the oil recirculating line to remove metal particles.

REFERENCES

*1 G. Boothroyd, *Fundamentals of Metal Machining and Machine Tools,* Scripta/McGraw-Hill, Washington, D. C., 1975.

 2 *Machining Data Handbook,* 2nd ed., Machinability Data Center, Metcut Research Associates, Cincinnati, OH, 1972.

 3 C. Miller, M. David, A. Goldman, and J. Wyatt, *AMA Arch. Ind. Hyg. Occup. Med.,* **8,** 453 (1953).

 4 J. A. Schey, *Introduction To Manufacturing Processes,* McGraw-Hill, New York, 1977.

 5 R. K. Springborn, Ed., *Cutting and Grinding Fluids: Selection and Application,* American Society of Tool and Manufacturing Engineers, Dearborn, MI, 1967.

 6 M. Key, E. Ritter, and K. Arndt. *Am. Ind. Hyg. Assoc. J.,* **27,** 423 (1966).

 7 "This is About Health, Esso Lubricating Oils and Cutting Fluids," Esso Petroleum Company, Ltd., London, 1979.

 8 D. O'Brien and J. C. Frede, "Guidelines for the Control of Exposure to Metalworking Fluids," Department of Health, Education and Welfare, Publication No. (NIOSH) 78-165, Cincinnati, OH, 1978.

 9 O. Einarsson, E. Eriksson, G. Lindstedt, and J. E. Wahlberg. *Contact Dermatitis J.,* **5,** 129 (1979).

10 H. A. Waldron, *J. Soc. Occup. Med.,* **27,** 45 (1977).

11 Committee on Industrial Hygiene, American Conference of Governmental Industrial Hygienists, *Industrial Ventilation: A Manual of Recommended Practice,* 16th ed., ACGIH, Lansing, MI, 1980.

12 "Nitrosamines in Cutting Fluids," NIOSH Current Intelligence Bulletin, Rockville, MD, 1976.

13 "ECM, ECD, ECG, Simplified," Chemform Division of KMS Industries, Pompano Beach, FL, 1970.

14 J. A. McGeough, *Principles of Electrochemical Machining,* Chapman and Hall, London, 1974.

2.11 METAL THERMAL SPRAYING

If a fine mesh metal powder or metal wire is melted in an application gun and the molten particles are conveyed to a workpiece at high velocity, they will freeze and adhere to the workpiece surface. A practical technique for spraying metal was invented in the 1920s. This technique has been used to apply metals and alloys to a workpiece for corrosion protection, to build up worn or corroded parts, to improve wear resistance, to reduce production costs, and to make decorative surfaces. In recent years metal alloys, ceramics, cements, and plastics have been sprayed as shown in Table 2.11–1 (1, 2). The health hazards from thermal spraying of metals were first observed in 1922 in the United Kingdom when operators suffered from lead poisoning (3).

The application head for thermal spraying takes many forms and dictates the range of hazards one may encounter from the operation. The four techniques presently in use in industry are described in the following sections.

Table 2.11–1 *Sprayed Materials*

Metals	Tungsten, nickel, chromium, tantalum, aluminum, zirconium, iridium, cadmium, lead, molybdenum
Alloys	Nichrome, stainless steel, babbitt, bronze
Ceramics	Alumina, zirconia, beryllia, spinel, zircon, glass
Intermetallic compounds	Nickel aluminide, molybdenum disulfide
Interstitial compounds	Titanium carbide, zirconium diboride, chromium carbide

Source.   Reference 1.

2.11.1 Spraying Methods

Combustion spraying

The coating material in wire form is fed to a gun fueled by a combustible gas such as acetylene, propane, or natural gas (Figure 2.11–1). The wire is melted in the oxygen-fuel flame, atomized with compressed air, and propelled from the torch at velocities up to 120 m/s (24,000 fpm). The material bonds to the workpiece by a combination of mechnical interlocking of the molten particles and a cementation of partially oxidized material.

The material can also be sprayed in powder form. The fuel gases in this system are either acetylene or hydrogen and oxygen. The powder is aspirated by an air stream and the molten particles are deposited on the workpiece with high efficiency.

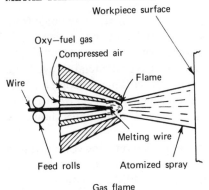

Gas flame

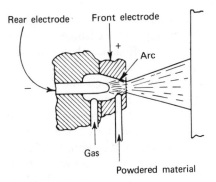

Figure 2.11-1 Combustion spray.   **Figure 2.11-2** Plasma spray.

Detonation spraying

In this unit the gases are fed to a combustion chamber where they are ignited by a spark plug at the rate of 260 firings per minute. Metal power is fed to the chamber, and the explosions drive the molten powder to the workpiece at velocities of approximately 760 m/s (150,000 fpm).

Plasma spraying

An electric arc is established in the controlled atmosphere of a special nozzle as shown in Figure 2.11–2. Argon is passed through the arc where it ionizes to form a plasma that continues through the nozzle and recombines to create temperatures as high as 16,700°C (30,000°F). Powder is melted in the stream and released from the gun at a velocity of approximately 10 m/s (2000 fpm).

Electric arc spraying

The consumable electrodes in this process are wire or metal to be sprayed. Two wires establish an arc as in a conventional arc welding unit and the molten metal is atomized by compressed air and projected to the workpiece at high velocity (Figure 2.11–3).

2.11.2 Hazards and Controls

The major hazard common to all techniques is the potential exposure to toxic metal fumes. In the gas torch spraying technique the particle size of molten metal conveyed to the part is stated to be approximately 50 μm (3).

If the vapor pressure of the metal at the application temperature is high, this will be reflected in high air concentrations of metal. If the metal is overheated due to malfunction or poor adjustment of the gun, this problem will be accentuated. The deposition efficiency, that is, the percentage of the metal sprayed deposited on the workpiece, varies with application techniques and coating metals. For flame metallizing with wire, the overspray varies from 11%

for aluminum to 45% for lead, while the plasma arc technique may range from 50% overspray for chromium carbide to as low as 10% for alumina-titania (2). Deposition efficiency obviously has a major impact on air concentrations in the workplace. Since the airborne level will vary with the type of metal sprayed, the application technique, the fuel in use, and the tool-workpiece geometry, it is impossible to estimate the level of contamination which may occur from a given operation. For this reason, if a toxic metal is sprayed, air samples should be taken to define the worker exposure.

Few industrial hygiene studies have been published on fume concentration during spray metallizing so it is difficult to characterize the potential hazards. However, the American Welding Society has published data on fume generation rates (4).

In addition to the airborne metal contaminant, the application techniques present the same range of potential health problems, including toxic gases, infrared and ultraviolet radiation, and noise, as comparable welding techniques (Section 2.15). The plasma spray presents the widest range of hazards, while combustion spraying presents the most limited.

One manufacturer has provided information on the noise level during operation of various equipment as shown in Table 2.11–2 (5). Appreciable noise reduction can be achieved by changing the operating characteristics of the spray equipment. On wire and powder guns the noise level can be reduced by 3 to 5 dBA by lowering the gas flow rate. Reduced amperage on plasma flow spray guns and electric arc guns is effective in reducing noise output from this equipment. However, there is a penalty for this noise reduction in lower spray rate and a poorer quality surface coating. If one cannot accept these penalties, the manufacturer suggests the normal approach to noise control including isolation, the use of hearing protectors, and various work practices.

The principal control of air contaminants generated during flame and arc spraying is local exhaust ventilation (6, 7). Nontoxic metals may be exhausted with a freely suspended open hood with a face velocity of 1.0 m/s (200 fpm) or an enclosing hood with a minimum face velocity of 0.65 m/s (125 fpm). Toxic

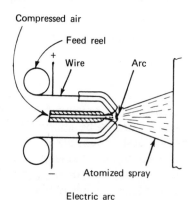

Electric arc

Figure 2.11-3 Electric arc spray.

Table 2.11-2 *Typical Noise Levels From Flame Spray Equipment*

Equipment	Operation	Noise Level Range (dBA)
Gas torch	Oxygen-acetylene	88–89
Powder spray gun	Acetylene	88–89
	Acetylene with air jet at 50 psi	110
	Hydrogen	100–101
Wire combustion gun	Acetylene	114
(1/8 and 3/16 in. wire)	Propane	118
	Methylacetylene propadiene	118
Electric arc guns	24 V, 200 A	111
	32 V, 500 A	116
Plasma flame spray gun	Nitrogen 125/0 flows, 600 A	134
	Nitrogen/hydrogen 75/15 flows, 600 A	133
	Argon 200/0 flows, 1000 A	128
	260/0 flows, 800 A	131
	Argon/hydrogen 80/15 flows, 600 A	132
	150/5 flows, 600 A	127

Source. Reference 5.

metal fumes should be exhausted with an enclosed hood with a face velocity of 1.0 m/s (200 fpm). The operator should wear an air-supplied respirator while metallizing with toxic metals. One must also consider the exposure of other workers in the area. Systems with 50% overspray obviously represent a significant source of general contamination.

In addition to air-supplied respirators, operators handling toxic metals such as cadmium and lead should wear protective gloves and coveralls and be required to strip, bathe, and change to clean clothes before leaving the plant.

Two special problems are encountered in routine metallizing operations. Unless ventilation control is effective, particles will deposit on various plant structures and may become a potential fire hazard. If one uses a scrubber to collect the particles, one may have hydrogen formed in the sludge, with a potential fire and explosion hazard.

REFERENCES

1 T. Lyman, Ed., *Metals Handbook,* 8th ed., Vol. 2, American Society for Metals, Metals Park, OH, 1964.

2 "The Metco Flame Spraying Processes," Metco Inc., Westbury, NY, 1978.

3 W. E. Ballard. *Ann. Occup. Hyg.,* **15**, 101 (1972).

4 American Welding Society, *Fumes and Gases in the Welding Environment,* AWS, Miami, FL, 1979.

5 Metco Technical Bulletin, "Flame Spray Noise Control, Occupational Safety and Health Act of 1970," Metco Inc., Westbury, NY, 1977.

6 J. H. Hagopian and E. K. Bastress, "Recommended Industrial Ventilation Guidelines," Department of Heath, Education and Welfare, Publication No. (NIOSH) 76–162, Cincinnati, OH, 1976.

*7 American Welding Society, *"Recommended Safe Practices for Thermal Spraying,"* AWS, Miami, FL, 1973.

2.12 NONDESTRUCTIVE TESTING

Many procedures are used in nondestructive testing in industry (1). Industrial radiography is covered separately in Section 2.9. This section deals with the common nondestructive techniques using dye penetrants and magnetic particles for inspection. These procedures have been widely used in industry for several decades without significant health hazards. Cleaning and preparation of materials for inspection by these methods may, however, present health hazards as described in other sections.

2.12.1 Penetrant Inspection

This technique is widely used to inspect for surface imperfections in any nonmagnetic solid. Its application is in the inspection of incoming material, such as bar stock, forgings, and billets, and for in-process and final inspection of parts.

Either fluorescent or visible dyes are suspended in a liquid carrier that is applied to the surface of the part by brushing, dipping, or spraying. If there are any surface imperfections, the penetrant is drawn into the opening by capillary action. The part is rinsed clear and a dry absorbent powder is applied by dusting or dipping. The penetrant in the imperfection "weeps" into the powder and is identified clearly by direct viewing if a visible dye is used or with a UV lamp in a dark room in the case of a fluorescent dye.

Diverse materials are used as dyes and carriers. Data are not available in the open literature on the composition of these materials. The dyes are commonly "taken up" in a low volatile oil and the absorbent is probably talc. In neither case do significant exposures occur; however, it is wise to obtain manufacturer's information on the specific materials in use in the system.

2.12.2 Magnetic Particle Inspection

This process complements the fluorescent penetrant system in that it is suitable for nondestructive testing of magnetic materials. The procedure is used for incoming, in-process, and final inspection.

In this process ferromagnetic particles are applied to the surface of a magnetic material by an air powder gun or by dipping in a bath containing the particles suspended in a light petroleum oil or water. If the material or part is then subjected to an induced magnetic field by a low voltage, high current power supply (2–16 V, 100–1000 A), the particles will be attracted to any surface discontinuity and will deposit on its edges, thereby defining its geometry. If the magnetic particles are designed to fluoresce, they can be identified by UV irradiation.

The particles may be suspended in water or a high flash point oil. The exact composition of the particles and the range of carrier liquids are not available in the open literature and should be requested from the manufacturer.

Although no comprehensive survey has been published, in the author's experience the health hazards from both of these operations are minimal. Dipping rather than spray application of the particles should be encouraged, and where possible an aqueous or low volatility carrier is preferred to minimize exposure. The UV irradiation is conducted with a low energy lamp and is not thought to present a hazard.

REFERENCES

1   American Welding Society, *Welding Handbook*, AWS Miami, FL, 1971.

2.13   PAINTING

Paint products are used widely in industry to provide a surface coating for protection against corrosion, for appearance, as electrical insulation, and for a number of special purposes. The hazards associated with the industrial application of these products will be discussed in this section. The manufacture of paint products will be covered in Chapter 3.

2.13.1   Types of Paints

The term paint is commonly used to identify a range of products including conventional paints, varnishes, enamels, and lacquers. Conventional paint is an inorganic pigment dispersed in a vehicle consisting of a binder and a solvent, with selected fillers and additives. Varnish is a nonpigmented product based on oil and resin in a solvent that dries first by the evaporation of the solvent and then by the oxidation and polymerization of the resin binder. A pigmented varnish is called an enamel. Lacquers are coatings that are commonly based on a cellulose ester in a solvent that dries by evaporation leaving a film that can be redissolved in the original solvent (1).

2.13.2   Composition of Paint

Solvent-based paints consist of the vehicle, filler, and additives as shown in Table 2.13–1 (2). The vehicle represents the total liquid content of the paint and includes the binder and the solvent. The binder may be a naturally occurring oil or resin or a synthetic material. In addition to linseed oil and oleoresinous materials, the common resins are listed in Table 2.13–2.

The fillers include pigments that historically have presented a major hazard in painting. The use of common heavy metal compounds including the lead

Table 2.13-1   *Paint Constituents*

Major Component	Constituents	Purpose
Vehicle	Binder	Resin that forms film
	Solvent	Thinner for adjustment of viscosity
Filler	General filler	Hiding ability, body, color
	Pigment	Opaqueness, color
	Extender	Fillers that build body
Additives	Driers	Speed drying or curing
	Biocide	Prevent growth of mold and fungus
	Flatting agents	Provide low luster
	Stabilizers	Protect against heat and UV radiation
	Antiskinning	Prevent skin formation in can

Source.   Reference 2.

Table 2.13-2   *Common Resins in Paint Systems*

Resin	Basis of Manufacture
Alkyd	Reaction of a dicarboxylic acid with polyols. May react alkyd resin with soybean oil, linseed oil, or phenolic resin
Modified alkyd	Reaction of alkyd resin with styrene, vinyl chloride, allyl alcohol, or urea-formaldehyde resin
Phenolic	Reaction of phenol or substituted phenols with an aldehyde to produce acidic or alkaline resins
Acrylic	Polymerization of esters of acrylic and methacrylic acids
Vinyl	Vinyl polymers and copolymers
Amino	Reaction of amino resins of urea and melamine with formaldehyde
Epoxy	Reaction of a phenol (Bisphenol A) with an epoxy (epichlorohydrin). Frequently mixed with amino or phenolic resins
Polystyrene	Styrene in ethyl benzene with catalyst (benzoyl peroxide)
Polyurethanes	Polyurethane prepolymer reacted with alcohols and amines
Cellulose	Esters or ethers of cellulose such as nitrocellulose

carbonates, sulfates, and oxides, cadmium compounds, and chromates have been curtailed; however, they are still present in certain paints as shown in Table 2.13–3. The use of lead pigments in protective paint has been reduced significantly with the introduction of low hazard pigments such as zinc phosphate. The solvent systems used in conventional paints may be chosen from a wide range of solvents as indicated in Table 2.13–4.

Table 2.13-3 *Pigments and Extenders*

White Pigments	Blue Pigments	Metallics and miscellaneous types
White lead—basic carbonate	Iron blue	Aluminum powders
White lead—basic sulfate	Ultramarine blue	Bronze powders
White lead—basic silicate	Indanthrene blue	Zinc dust
Zinc oxide	Others	Lead powder and flake
Leaded zinc oxide	**Red Pigments**	Nickel flake
Titanium dioxides—anatase and rutile	Cadmium red	Stainless steel
Lithopone	Cadmium-mercury reds and oranges	Cuprous oxide
Antimony oxide	Toluidine red	Others
Iron Oxide	Para red	**Extenders**
Red and brown iron oxides—natural	Toners and others	Barytes
Red and brown iron oxides—synthetic	**Yellow and Orange Pigments**	Blanc fixe
Yellow iron oxide—synthetic	Chrome yellow	Calcium carbonate
Ochre	Chrome orange	Silica—amorphous
Raw and burnt siennas	Molybdate orange	Silica—diatomaceous
Raw and burnt umbers	Zinc yellow	Talc
Black iron oxide	Cadmium yellow and orange	China clay
Green Pigments	Hansa yellow	Mica
Chrome green	Orange toner	Bentonite
Chromium oxide	**Luminous and Fluorescent Pigments**	Asbestos
Phthalocyanine green	Several colors	Others
Others		

Source. Reference 2. From *Surface Preparation and Finishes for Metals* by J. Murphy. Copyright 1971. Used with permission of McGraw-Hill Book Co.

Table 2.13-4 *Partial List of Paint Solvents—Thinners*

Aromatic	Chlorinated Solvents	Acetates
Benzene	Methyl chloride	Ethyl
Toluene	Chlorothene	Isopropyl
Xylene	Carbon tetrachloride	*n*-Propyl
Aromatic naphthas	Ethylene dichloride	Secondary butyl
Aromatic petroleum	Trichloroethylene	*n*-Butyl
solvents	Perchloroethylene	Amyl
Others	**Terpenes**	**Ketones**
Aliphatic	Turpentine	Acetone
Petroleum ether	Dipentene	Methyl ethyl ketone
Lacquer diluent	Pine oil	Methyl acetone
VM and P naphtha	**Alcohols**	Methyl isobutyl ketone
Mineral spirits	Methanol	Diacetone
Odorless mineral spirits	Ethanol	Cyclohexanone
Kerosene	Isopropyl alcohol	Isophorone
High flash naphthas	*n*-Propyl alcohol	Diisobutyl ketone
Others	*n*-Butyl alcohol	
Glycol Ethers	Secondary butyl alcohol	
Several commercial	Amyl alcohol	
grades	Cyclohexanol	
	Others	

Source. Reference 2. From *Surface Preparation and Finishes for Metals* by J. Murphy. Copyright 1971. Used with permission of McGraw-Hill Book Co.

In addition to the solvent-based paints, water-thinnable paints are available and represent a significant portion of the consumer market. The first water-based paints, or latex paints, utilized an emulsion of styrene-butadiene rubber. The present systems commonly are based on an emulsion of polyvinyl acetate or acrylic resins either as a polymer or copolymer. These paints are based on water-soluble polymers that become water-insoluble after application. A number of the resins noted in Table 2.13-2 can be made soluble by joining a carboxylic, hydroxyl, epoxy, or amine group of the polymer with coupling agents such as alcohols or glycol ethers.

Special attention must be given epoxy and urethane paint systems. The two component urethane paint systems consist of a polyurethane prepolymer containing a reactive isocyanate and a polyester. Mixing of the two materials initiates a reaction with the formation of a chemically resistant coating. In the first formulations the unreacted toluene diisocyanate (TDI) resulted in significant airborne exposures to TDI and resulting respiratory sensitization. Formulations have been changed to include TDI adducts and derivatives that are stated to have minimized the respiratory hazard from these systems. For ease of application, single package systems have been developed using a blocked isocyanate. Unblocking takes place when the finish is baked and the isocyanate is released to react with the polyester.

Epoxy paint systems offer excellent adhesive properties, resistance to abrasion and chemicals, and stability at high temperatures. The conventional epoxy system is a two component system consisting of a resin based on the reaction products of bisphenol A and epichlorohydrin. The resin may be modified by reactive diluents such as glycidyl ethers. The second component, the hardener or curing agent, was based initially on low molecular weight, highly reactive amines (3).

Lemon states that the epoxy resins can be divided into three grades (4). The solid grades are felt to be innocuous; however, skin irritation may occur from solvents used to take up the resin. The liquid grades are mild to moderate skin irritants. The low viscosity glycidyl ether modifiers are skin irritants and sensitizers and have systemic toxicity. The use of low molecular weight aliphatic amines such as diethylene tetramine and triethylene tetramine, strong skin irritants and sensitizers, presented major health hazards during the early use of epoxy systems. According to Lemon, this has been overcome to a degree by the use of low volatility amine adducts and high molecular weight amine curing agents.

2.13.3 Operations and Exposures

In the industrial setting, paints can be applied to parts by a myriad of processes including brush, roller, dip, flow, curtain, tumbling, conventional air spray, airless spray, steam atomization, disc spraying, and powder coating. Conventional air spraying is the most common method encountered in industry and presents the principal hazards due to overspray and rebound. The use of airless and hot spray techniques minimizes mist and solvent exposure to the operator, as does electrostatic spraying. The electrostatic technique, now commonly used with many spray atomization installations, places a charge on the paint mist particle so that it is attracted to the part to be painted.

The operator is exposed to the solvent or thinner in processes in which the paint is flowed on as in brushing, in dipping, and during drying of the parts. However, during atomization techniques the exposure is to both the solvent and the paint mist. The level of exposure reflects the overspray and rebound that occurs during spraying as shown in Table 2.13-5 (5).

Table 2.13-5 *Overspray During Painting*

Method of Spraying	Percentage Overspray	
	Flat Surface	Table Leg
Air atomization	50	85
Airless	20–25	90
Electrostatic disc	5	5–10
Airless	20	30
Air atomization	25	35

Source. Reference 3.

A common exposure to organic vapors occurs when the spray operator places the freshly sprayed parts on a rack directly behind him. The air movement to the spray booth sweeps over the drying parts and past the breathing zone of the operator resulting in an exposure to solvent vapors. Drying stations and baking ovens must, therefore, be exhausted. The choice of exhaust control on tumbling and roll applications depends on the surface area of the parts and the nature of the solvent.

2.13.4 Controls

Most industrial flow and spray painting operations utilizing solvent-based paints require exhaust ventilation for control of solvent vapors at the point of application and during drying and baking operations. Water-based paints may only require ventilation when spray application is utilized. Flow application of solvent-based paints will require local exhaust ventilation depending on the application technique (6). The usual ventilation controls for spray application of solvent-based paints are in the form of a spray booth, room, or tunnel provided with some type of paint spray mist arrestor before it is exhausted outdoors. The degree of control of paint mist and solvent varies with the application method, that is, whether it is air atomization, airless, or electrostatic painting. The latter two techniques call for somewhat lower exhaust volumes.

The design of the conventional paint spray booth is shown in Figure 2.13-1 (6, 7). The booths are commonly equipped with a water curtain or a throwaway dry filler to provide paint mist removal. The efficiency of these devices against paint mist has not been evaluated. Neither of these systems, of course, removes solvent vapors from the air stream.

In industrial spray painting of parts, the simple instructions contained in several state codes on spray painting should be observed:

1 Do not spray toward a person.
2 Automate spray booth operations where possible to reduce exposure.
3 Maintain 0.6 m (2 ft) clearance between sides of the booth and large flat surfaces to be sprayed.
4 Keep the distance between the nozzle and the part to be sprayed less than 0.3 m (1 ft).

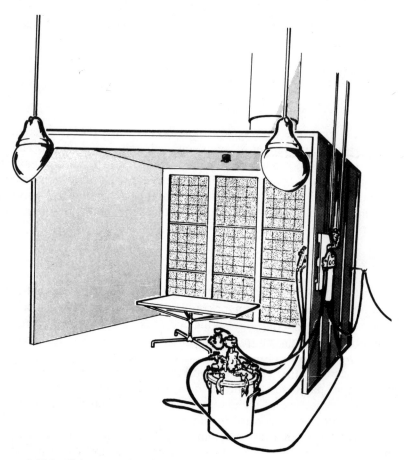

Figure 2.13-1 Paint spray booth.

5 Do not position work so that the operator is between the exhaust and the spray gun or disc.

6 Locate the drying room so that air does not pass over drying objects to the exhaust hood past the breathing zone of the operator.

Controls in the application of two component urethane and epoxy paint systems must include excellent housekeeping, effective ventilation control, and protective clothing; in applications not effectively controlled by ventilation, the operators should wear air-supplied respirators. Adequate washing facilities should be available, and eating, drinking, and smoking should be prohibited in the work area.

Dermatitis due to primary irritation and defatting from solvents or thinners, as well as sensitization from epoxy systems, is not uncommon. Skin contact must be minimized, rigorous personal cleanliness encouraged, and suitable protective equipment used by the operator.

Worker exposure during industrial painting of parts has not been documented, nor has exposure during painting of tanks, ships, and other enclosures been studied. An evaluation of the effectiveness of respiratory protection in the paint spraying industry involving approximately 2600 painters revealed that the observed respirator protection factors determined by workplace sampling did not approach the commonly accepted protection factors due to poor selection, inadequate training, and lack of maintenance (8).

REFERENCES

1 "Standard Definitions of Terms Relating to Paint, Varnish, Lacquer, and Related Products," ASTM Designation D16 (1967) ASTM Standards, Part 20, American Society for Testing Materials, Philadelphia, PA, 1968.

*2 J. Murphy Ed., *Surface Preparation and Finishes for Metals,* McGraw-Hill, New York, 1971

3 R. Piper, *Br. J. Ind. Med., 22,* 247 (1965).

4 R. Lemon, *Ann. Occup. Hyg., 15,* 131 (1972).

5 J. Danielson, Ed., *Air Pollution Engineering Manual,* 2nd ed., Publication No. AP-40, Government Printing Office, Washington, D.C., 1973.

6 Committee on Industrial Ventilation, American Conference of Government Industrial Hygienists, *Industrial Ventilation: A Manual of Recommended Practice,* 16th ed., ACGIH, Lansing, MI, 1980.

7 American National Standards Institute, "American Standard Safety Code for the Design, Construction, and Ventilation of Spray Finishing Operations," A9.3-1964, ANSI, New York, reaffirmed, 1971.

8 C. R. Toney and W. L. Barnhart, "Performance Evaluation of Respiratory Protective Equipment Used in Paint Spraying Operations," U.S. Department of Health, Education and Welfare, Publication No. (NIOSH) 76-177, Cincinnati, OH, 1976.

2.14   SOLDERING AND BRAZING

Soft soldering is the joining of metal using filler metal (solder) with a melting point less than 316°C (600°F); hard solder is used in the range of 316–427°C (600–800°F). This differentiates soldering from brazing, which utilizes a filler metal with a melting point greater than 427°C (800°F).

2.14.1   Soldering

To understand the potential health hazards from soldering operations, one must be familiar with the composition of the solder and fluxes in use and the applicable production techniques.

Flux

All metals, including the noble metals, have a film of tarnish that must be removed in order to effectively wet the metal with solder and accomplish a good mechanical bond. The tarnish takes the form of oxides, sulfides, carbonates, and other corrosion products (1). The flux, which may be a solid, liquid, or gas, is designed to remove any adsorbed gases and tarnish from the surface of the base metal and keep it clean until the solder is applied. The molten solder displaces the residual flux and wets the base metal to accomplish the bond.

Diverse organic and inorganic materials are used in the design of soldering fluxes, as shown in Table 2.14-1. The flux is usually a corrosive cleaner frequently used with a volatile solvent or vehicle. Rosin, which is a common base for organic fluxes, contains abietic acid as the active material. Manko has described the performance of this flux on a copper base metal that has an oxidized surface (1). When the flux is applied, the surface oxide reacts with the abietic acid to form copper-abiet. The copper-abiet mixes with the unreacted rosin on the now untarnished surface of the base metal. When the molten solder is applied, it displaces the copper-abiet and the unreacted rosin to reveal the clean metal surface, thereby ensuring a good bond. Activated rosin flux, formulated by adding an organic chloride compound to the base rosin-alcohol liquid flux, provides active corrosive cleaning. The three major flux bases are inorganic, organic-nonrosin, and organic-rosin, as shown in Table 2.14-1. A given flux may contain several of these materials. The inorganic fluxes may be based on any of the acids commonly used for etch baths in electroplating operations. The inorganic salts, such as zinc chloride, react with water to form hydrochloric acid, which converts the base metal oxide to a chloride. The metal chloride is water soluble and can be removed easily by a water rinse to provide an untarnished surface for soldering. A combination of salts is frequently used to achieve a flux with a melting point below that of the solder so that no flux residual will be left on the part. The last inorganic technique utilizes hot hydrogen or hydrogen chloride gas, which sweeps over the base metal reducing the oxide.

Table 2.14-1 *Common Flux Materials*

Type	Typical fluxes	Vehicle
	Inorganic	
Acids	Hydrochloric, hydrofluoric, orthophosphoric	Water, petrolatum paste
Salts	Zinc chloride, ammonium chloride, tin chloride	Water, petrolatum paste, polyethylene glycol
Gases	Hydrogen forming gas; dry HCl	None
	Organic; Nonrosin Base	
Acids	Lactic, oleic, stearic, glutamic, phthalic	Water, organic solvents, petrolatum paste, polyethylene glycol
Halogens	Aniline hydrochloride, glutamic acid hydrochloride, bromide derivatives of palmitic acid, hydrazine hydrochloride or hydrobromide	Water, organic solvents, polyethylene glycol
Amines and amides	Urea, ethylenediamine, mono- and triethanolamine	Water, organic solvents, petrolatum paste, polyethylene glycol
	Organic: Rosin Base	
Superactivated	Rosin or resin with strong activators	Alcohols, organic solvents, glycols
Activated (RA)	Rosin or resin with activator	Alcohols, organic solvents, glycols
Mildly activated (RMA)	Rosin with activator	Alcohols, organic solvents, glycols
Nonactivated (water-white rosin) (R)	Rosin only	Alcohols, organic solvents, glycols

Source.   Reference 1.

The second class of materials shown in Table 2.14-1, organic nonrosin, is less corrosive and therefore slower acting. In general, these fluxes do not present as severe a handling hazard as the inorganic acids. This is not true of the organic halides whose degradation products are very corrosive and warrant careful handling and ventilation control. The amines and amides in this second class of flux materials are common ingredients whose degradation products are very corrosive.

The rosin base fluxes listed in Table 2.14-1 are inactive at room temperature, but at soldering temperature they are activated to remove tarnish. This material was and still is a popular flux compound since the residual material is chemically and electrically benign and need not be removed as vigorously as the residual products of the corrosive fluxes.

The cleaning of the soldered parts may range from a simple hot water or detergent rinse to a degreasing technique using a range of organic solvents. These cleaning techniques may require local exhaust ventilation.

Solder

The most common solder contains 65% tin and 35% lead. Traces of other metals, including bismuth, copper, iron, aluminum, zinc, and arsenic, are present. Several special solders contain antimony in concentrations up to 5%. The melting point of these solders is quite low and at these temperatures the vapor pressure of lead and antimony usually do not result in significant air concentrations of metal fume. The composition of the common industrial solders is shown in Table 2.14-2.

Application techniques

Soldering is a fastening technique used in a wide range of products from simple mechanical assemblies to complex electronic systems. The soldering process includes cleaning the base metal and other components for soldering, fluxing, the actual soldering and postsoldering cleaning.

Initial cleaning of base metals

Prior to fluxing and soldering, the base metal must be cleaned to remove oil, grease, wax, and other surface debris. Unless this is done, the flux will not be able to attack and remove the metal surface tarnish. The procedures used in metal cleaning, reviewed in Section 2.3, include cold solvent degreasing, vapor degreasing, and ultrasonic degreasing. If the base metal has been heat treated, the resulting surface scale must be removed by one of the procedures described in Section 2.2. In many cases mild abrasive blasting techniques are utilized to remove heavy tarnish before fluxing (Section 2.1). In electrical soldering the insulation on the wire must be stripped back to permit soldering. Stripping is accomplished by mechanical techniques using cutters or wire brushes, chemical strippers, or thermal techniques. Mechanical stripping of asbestos-based insulation obviously presents a potential health hazard and this operation must be controlled by local exhaust ventilation. The hazard from chemical stripping depends on the chemical used to strip the insulation; however, at a minimum the stripper will be a very corrosive agent. Thermal stripping of wire at high production rates may present a problem due to the thermal degradation products of the insulation. Hot wire stripping of fluorocarbon insulation such as Teflon® and Terzel® may cause polymer fume fever if operations are not controlled by ventilation. Other plastic insulation such as PVC (polyvinyl chloride) may produce irritating and toxic thermal degradation products. A summary of the common stripping techniques for various insulation materials is shown in Table 2.14-3.

Table 2.14-2 *Typical Solder Compositions*

ASTM Classification	Composition (wt. %)		Temperature (°F) Solidus Liquidus Pasty Range		
	Tin	Lead			
5A	5	95	578	594	36
10A	10	90	514	573	59
15A	15	85	437	553	116
20A	20	80	361	535	174
25A	25	75	361	511	150
30A	30	70	361	491	130
35A	35	65	361	477	116
40A	40	60	361	455	94
45A	45	55	361	441	80
50A	50	50	361	421	60
60A	60	40	361	374	13
70A	70	30	361	373	17
	Tin	Antimony	Lead		
20C	20	1.0	79.0	363	517	154
25C	25	1.3	73.7	364	504	140
30C	30	1.6	68.4	364	482	118
35C	35	1.8	63.2	365	470	105
40C	40	2.0	58.0	365	448	83
95TA	95	5.0	58.0	452	464	12
	Lead	Silver	Tin		
2.5S	97.5	2.5	—	579	579	0
5.5S	94.5	5.5	—	579	689	110
1.5S	97.5	1.5	1.0	588	588	0
	Tin	Silver			
	96.5	3.5	430	430	0
	95	5	430	473	43
	Tin	Zinc			
	91	9	390	390	0
	80	20	390	518	128
	70	30	390	592	202
	60	40	390	645	255
	30	70	390	708	318
	Cadmium	Zinc			
	82.5	17.5	509	509	0

Table 2.14-2 *Typical Solder Compositions —Continued*

ASTM Classification		Composition (wt. %)			Temperature (°F) Solidus	Liquidus	Pasty Range
		Cadmium	**Zinc**				
		40	60		509	635	126
		10	90		509	751	241
Tin	**Indium**	**Bismuth**	**Lead**	**Cadmium**			
8.3	19.1	44.7	22.6	5.3	117	117	0
12	21	49	18	—	136	136	0
12.8	4	48	25.6	9.6	142	149	7
50	50	—	—	—	243	257	14
—	50	—	50	—	356	408	52

Source.   Reference 3.

Fluxing operations

Proprietary fluxes are available in solid, paste, and liquid form for various applications and may be cut with volatile vehicles such as alcohols to vary their viscosity. The flux may be applied by one of the 10 techniques shown in Table 2.14-4 depending on the component and the production rate. Since one is handling a corrosive chemical, skin contact must be minimized by specific work practices and good housekeeping. The two techniques that utilize spray application of the flux require local exhaust ventilation to prevent air contamination from the flux mist.

Soldering and cleaning

In two of the fluxing techniques listed in Table 2.14-4, the flux and solder are applied together. One must merely apply the necessary heat to bring the system first to the melting point of the flux and then to the melting point of the solder. In most applications, however, the flux and solder are applied separately, although the two operations may be closely integrated and frequently completed in a continuous operation.

The soldering techniques used for most manual soldering operations on parts that have been fluxed are the soldering iron and the solder pot techniques. A number of variations of the solder pot have been introduced to handle high production soldering of printed circuit boards (PCB) in the electronic industry. In the drag solder technique, the PCB is positioned horizontally and pulled along the surface of a shallow molten solder bath behind a skimmer plate that removes the dross. This process is usually integrated with a cleaning and fluxing station in a single automated unit. Another system designed for automation is wave soldering. In this technique a standing wave is formed by pumping the solder through a spout. Again, the conveyorized PCBs are pulled through the flowing solder.

The potential health hazards to the soldering operators are minimal. A simple lead-tin solder is used routinely at temperatures that have been considered too low to generate significant concentration of lead fume. This stance should be reconsidered based on the present OSHA Permissible Exposure Limit (PEL) of 50 μg/m³. The use of activated rosin fluxing agents may result in the release of thermal degradation products that require control by local exhaust ventilation. The handling of solder dross during cleanup and maintenance may result in exposure to lead dust.

Table 2.14-3 *Wire Insulation-Stripping Methods*[a]

Insulation Material	Mechanical	Thermal	Chemical
Asbestos ✱	1	3	3
Cloth	1	2x	3
Natural rubber	1	2x	3
Neoprene	1	2x	3
Nylon	1	2x	3
Paper	1	2x	3
Polyurethane	2	1	1
PVC (polyvinyl chloride)	1	2x	3
Rolan[b]	1	3	3
Silicon rubber	1	3	3
Solder Eaze[b]	2	1	2
Teflon[bc]	1	1 ✱	3
Varnish	2	1	1

Source. Reference 1. From *Solders and Soldering* by H. H. Manko. Copyright 1979. Used with permission of McGraw-Hill Book Co.

[a] 1 = normally used; 2 = used only under special conditions; 3 = not normally used; x = used mostly to separate unwanted sleeve from desired part of insulation; the rest of strip is mechanical.
[b] Trade name.
[c] Provide adequate ventilation for thermal strip.

Table 2.14-4 *Flux Application*

Method	Application Technique	Use	Solid	Paste	Liquid
Brushing	Applied manually by paint, acid, or rotary brushes	Copper pipe, job shop, PCB, large structural parts		X	X
Rolling	Paint roller application	Precision soldering, PCB, suitable for automation		X	X
Spraying	Spray painting equipment	Automatic soldering operations, not effective for selective application			X
Rotary screen	Liquid flux picked up by screen and air directs it to part	PCB application			X

Table 2.14-4   *Flux Application—Continued*

Method	Application Technique	Use	Solid	Paste	Liquid
Foaming	Work passes over air agitated foam at surface of flux tank	Selective fluxing, automatic PCB lines			X
Dipping	Simple dip tank	Wide application for manual and automatic operations on all parts		X	X
Wave	Liquid flux pumped through trough forming wave through which work is dipped	High speed, automated operation			X
Floating	Solid flux on surface melts providing liquid layer	Tinning of wire and strips of material	X		
Cored solder	Flux inside solder wire melts, flows to surface, and fluxes before solder melts	Wide range of manual operations	X	X	X
Solder paste	Solder blended with flux. Applied manually	Component and hybrid microelectric soldering		X	

Source.   Reference 1. From *Solders and Soldering* by H. H. Manko. Copyright 1979. Used with permission of McGraw-Hill Book Co.

Furnace soldering is used widely for the assembly of semiconductors. The parts are positioned in a holding jig and flux and solder preforms are positioned at the soldering locations. Parts are processed by batch or continuous furnace operations. Since small quantities of flux and solder are used at closely regulated furnace temperatures, there is no significant health hazard from this operation. In another operation, the parts are prefluxed and solder preforms are processed through a fluorocarbon vapor phase unit, which brings the assembly up to a precise temperature in a gradual fashion without undue thermal stress to the part. As in the case of the vapor degreaser, the condenser and freeboard ensure minimal release of the fluorocarbon to the workplace, but this should be evaluated.

After soldering operations are carried out, some flux residue and its degradation products remain on the base metal. Both water-soluble and solvent-soluble materials normally exist, so that it is necessary to clean with both systems using detergents and saponifiers in one case and common chlorinated and fluorinated hydrocarbons in the other. The processing equipment may include ultrasonic cleaners and vapor degreasers.

Controls

The fluxes used in soldering may represent the most significant potential hazard from soldering. The conventional pure rosin fluxes are not difficult to handle, but highly activated fluxes warrant special handling instruction. A range of alcohols, including methanol, ethanol, and isopropyl alcohol, are used as volatile vehicles for fluxes. The special fluxes should be evaluated under conditions of use to determine worker exposure.

2.14.2 Brazing

This technique of joining metals is discussed in this section since it is commonly identified as a soldering operation in industry. It is more properly identified as a welding process and is so classified by the American Welding Society. Brazing techniques are used widely in the manufacture of refrigerators, electronics, jewelry, and aerospace components to join both similar and dissimilar metals. Although the final joint looks similar to a soft solder bond, it is of much greater strength and the joint requires little finishing. As mentioned earlier in this section, brazing is defined as a technique for joining metals that are heated above 430°C (800°F), while soldering is conducted below that point. The temperature of the operation is of major importance since it determines the vapor pressure of the metals that are heated and therefore the concentration of metal fumes to which the operator is exposed.

Flux

Flux is frequently used; however, certain metals may be joined without flux. The flux is chosen to prevent oxidation of the base metal and not to prepare the surface, as is the case with soldering. The common fluxes are based on fluorine, chlorine, and phosphorous compounds, and they present the same health hazards as fluxes used in soldering: they are corrosive to the skin and can cause respiratory irritation. A range of filler metals are used as shown in Table 2.14-5 and include phosphorous, silver, zinc, copper, cadmium, nickel, chromium, beryllium, magnesium, and lithium. This selection of the proper filler metal is the key to quality brazing.

Application techniques

Brazing of small job lots that do not require close temperature control are routinely done with a torch. More critical, high production operations are accomplished by dip techniques in a molten bath, by brazing furnaces using either an ammonia or hydrogen atmosphere, or by induction heating (Table 2.14-5).

The brazing temperatures define the relative hazard from the various operations. For example, the melting point of cadmium is approximately 1400°C. The vapor pressure of cadmium and the resulting airborne fume concentrations increase dramatically with temperature as shown in Table 2.14-6. The filler metals with the higher brazing temperatures, therefore, will present the most severe exposure to cadmium. The exposure to fresh cadmium fumes during brazing of low alloy steels, stainless steels, and nickel alloys has resulted in documented cases of occupational disease and represents the major hazard from these operations. This is especially true of torch brazing, where temperature extremes may occur. On the other hand, the temperatures of furnace and induction heating operations may be controlled to ±5°C.

Table 2.14-5   *Application of Brazing*

Base Metal to Be Brazed	Brazing Filler Metals (AWS-ASTM designation)	Brazing Technique
Stainless steels	Silver alloys (BAg)	Torch
	Copper alloys (BCu, BCuP, BCuZN)	Furnace, dry hydrogen
Aluminum	Aluminum-silicon alloys (BAlSi)	Furnace
		Induction
		Dip
Magnesium	Magnesium alloys (BMg)	Torch
		Dip
Copper	Copper-phosphorus (BCuP)	All techniques
	Copper-zinc (BCuZn)	
	Silver alloys	
Nickel	Silver alloys (BAg)	All techniques
	Copper alloys (BCuP, BCuZn)	
Low alloy steels	Silver alloys (BAg)	All techniques
	Copper alloys (BCuP, BCuZn)	
	Nickel alloys (BNi)	

Table 2.14-6   *Vapor Pressure of Cadmium*

Temperature (°C)	Vapor Pressure (mm Hg)	Cadmium Concentration (mg/m³ at 25°C)
157	0.000021	0.12
200	0.00034	2.01
227	0.0015	8.9
321 (m.p.)	0.095	560
392	1.0	5900

Source.   Reference 4.

One study of brazing in a pipe shop and on board ship included air sampling of cadmium from brazing filler metal, fluorides from flux, and nitrogen dioxide from the torch (Table 2.14-7) (5). The cadmium content of the filler metal in these operations ranged from 10 to 24%. In reconstructing one accident aboard ship attributed to cadmium fume, the authors of this study found that the filler metal did not contain cadmium; however, the nitrogen dioxide concentrations reached 122 ppm in ½ hr of brazing and accounted for the disease pattern, which closely resembled that of exposure to fresh cadmium fume (5).

Table 2.14-7 *Airborne Exposure Levels During Silver Brazing[a]*

| Range | Pipe Shop Exposures | | |
	CdO (mg/m³)	F (mg/m³)	NO₂ (ppm)
High	0.014	0.16	4.0
Low	0.001	0.02	0.02
Mean	0.01	0.12	0.35
	Shipboard Exposures		
High	1.40	0.80	3.60[a]
Low	0.08	0.28	0.02
Mean	0.45	0.51	0.38

Source. Reference 5.

[a] Except for large piping when long heating cycles and large torches produce high concentrations of NO_2.

Controls

Controls from brazing operations must obviously be based on the identification of the composition of the filler metals using Table 2.14-5. Local ventilation control is necessary on operations where toxic metal fumes may be generated from the brazing components or from parts plated with cadmium or other toxic metals. In the use of common fluxes, one should minimize skin contact due to their corrosiveness and provide exhaust ventilation to control airborne thermal degradation products released during the brazing operation.

REFERENCES

*1 H. H. Manko, *Solders and Soldering,* McGraw-Hill, New York, 1979.

2 American Welding Society, *Soldering Manual,* AWS, Miami, FL, 1959.

3 Anonymous, *Machine Design,* **43,** 177, (Nov 1971).

4 H. Elkins, *Chemistry of Industrial Toxicology,* Wiley, New York, 1950.

5 C. Mangold and R. Beckett, *Am. Ind. Hyg. Assoc. J.,* **32,** 115, (1971).

2.15 WELDING

Welding is a process for joining metals in which coalescence is produced by heating the metals to a suitable temperature. The welding processes can be classed as pressure, nonpressure, or brazing (1). The nonpressure welding techniques warrant major attention because they involve the fusing or melting together of metals. Selected metal cutting operations are included in this discussion since these techniques are similar to the welding procedures and have comparable health hazards. Soldering, brazing, and metal thermal spraying are discussed in Sections 2.11 and 2.14.

In the nonpressure welding techniques, metal is vaporized and then condenses to form a fume in the 0.01 to 0.1 μm particle size range that agglomerates rapidly. The source of this respirable metal fume is the base metal, the metal coating on the workpiece, the electrode, and the fluxing agents associated with the particular welding system. A range of gases and vapors including ozone and nitrogen dioxide may be generated depending on the welding process. The wavelength and intensity of the electromagnetic radiation emitted depend on the welding procedure, inert gas, and base metal.

The common welding techniques will be reviewed assuming that the welding is done on mild steel. The nature of the base metal is important in evaluating the metal fume exposure and occasionally it has impact in other areas. When such impact is important it will be discussed. The information presented in the text is summarized in Table 2.15-1. The reports by the American Welding Society have been helpful in presenting this review and the reader is referred to these sources for more detail (2, 3).

Table 2.15-1 *Potential Hazards from Welding*[a]

Welding Process	SMA	Low Hydrogen	GTA[b]	GMA[c]	Submerged	Plasma	Resistance	Gas
Metal fumes	M–H	M–H	M–H	M–H	L	H	L	L–M
Fluorides	L	H	L	L	M	L	L	L
Ozone	L	L	M	H	L	H	L	L
Nitrogen dioxide	L	L	M	M	L	H	L	H
Carbon monoxide	L	L	L	L	L	M	L	M–H
				H if CO$_2$				
Decomposition of chlorinated HC	L	L	M	M–H	L	H	L	L
Radiant energy	M	M	M–H	M–H	L	H	L	L
Noise	L	L	L	L	L	H	L	L

[a] Hazard codes: L = low; M = medium; H = high.
[b] GTA = gas tungsten arc, [c] GMA = gas metal arc.

2.15.1 Shielded Metal Arc (SMA) Welding (Figure 2.15-1)

This most common nonpressure or fusion welding process is commonly called stick or electrode welding. An electric arc is drawn between a welding rod and the workpiece, melting the metal along a seam or a surface. The molten metal from the workpiece and the electrode form a common puddle and cool to form the weld and its slag cover (Figure 2.15-2). Power, dc or ac, is used in either straight (electrode negative, work positive) or reverse polarity. The most common technique involves dc voltages of 10 to 50 and a wide range of current

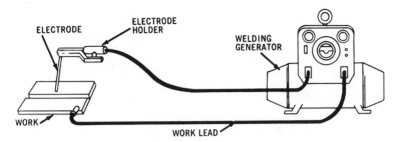

Figure 2.15-1 Shielded metal arc (SMA) welding. (Courtesy Hobart Brothers Co.)

up to 2000 A. Although operating voltages are low, under certain conditions an electrical hazard may exist.

The welding rod or electrode may have significant occupational health implications. Initially a bare electrode was used to establish the arc and act as filler metal. Now the electrode covering may contain 20 to 30 organic and inorganic compounds and perform several functions. The principal function of the electrode coating is to release a shielding gas such as carbon dioxide to insure that air does not enter the arc puddle and thereby cause failure of the weld. In addition, the covering stabilizes the arc, provides a flux and slag producer to remove oxygen from the weldment, adds alloying metal, and controls the viscosity of the metal.

Table 2.15-2 *Composition of Coverings on Representative Carbon Steel Electrodes*

	Covering Composition (wt. %)		
Constituent	E6013 High Cellulose —Sodium	E6013 High Titania —Potassium	E7016 Low Hydrogen —Potassium
SiO_2	32.0	25.9	16.0
$TiO_2 = ZrO_2$	18.0	30.6	6.5
Al_2O_3	2.0	5.9	1.0
CaF_2	—	—	27.0
CaO	—	1.6	—
MgO	6.0	2.6	—
Na_2O	8.0	1.1	1.4
K_2O	—	6.7	1.0
CO_2	—	1.7	—
Organics	30.0	17.7	—
Fe	2.0	2.1	—
Mn	7.0	4.8	2.5
$CaCO_3$	—	—	38.0

Source. Reference 3. Copyright 1979 American Welding Society. Reprinted by permission.

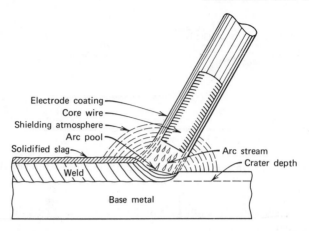

Figure 2.15-2 Weldment cross section.

A complete occupational health survey of shielded arc welding requires the identification of the rod and its covering. The range of compounds in these common electrode coverings is shown in Table 2.15-2. The composition of the electrode can be obtained from the AWS classification number stamped on the electrode as shown in Figure 2.15-3 (4).

The percent of total electrode mass that appears as airborne fume ranges from 0.5 to 5%. Pantucek states that the "fume" generated from coated electrodes may contain iron oxides, manganese oxide, fluorides, silicon dioxide, and compounds of titanium, nickel, chromium, molybdenum, vanadium, tungsten, copper, cobalt, lead, and zinc (5). The manganese content of welding fume ranges from 3 to 10%. Silicon dioxide is routinely reported as present in welding fume, apparently originating from the various silicon

FOURTH DIGIT	TYPE OF COATING	WELDING CURRENT
1	cellulose potassium	AC or DC Reverse or Straight
2	titania sodium	AC or DC Straight
3	titania potassium	AC or DC Straight or Reverse
4	iron powder titania	AC or DC Straight or Reverse
5	low hydrogen sodium	DC Reverse
6	low hydrogen potassium	AC or DC Reverse
7	iron powder iron oxide	AC or DC
8	iron powder low hydrogen	DC Reverse or Straight or AC

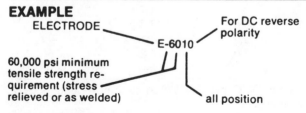

Figure 2.15-3 AWS electrode classification.

compounds in the electrode coating. Buck and Dessler state that the silicon dioxide is present in an amorphous form or as silicates, not in the crystalline form (6).

Metal fume exposure

The potential health hazards from exposure to metal fume during shielded metal arc welding obviously depend on the metal being welded and the composition of the welding electrode. This discussion will be directed to mild steel welding so that the principal component of the fume generated from the workpiece is iron oxide. The hazard from exposure to iron oxide fume appears to be limited. The fume does cause a benign pneumoconiosis known as siderosis due to the deposition of iron oxide particulate in the lung. There is no disability or fibrous tissue proliferation resulting from exposure to iron oxide fume; however, the possibility of pulmonary function decrement is in dispute.

The concentration of metal fume to which the welder is exposed depends on the welding conditions including the current density (amperes per unit area of electrode), the arc time, which may vary from 10 to 30% of the shift, the power configuration, that is, dc or ac supply, and straight or reverse polarity. The operating environment also defines the level of exposure to welding fume and includes the type and quality of exhaust ventilation and whether the welding is done in an open, enclosed, or confined space.

Much of the data on metal fume exposures have been generated from shipyard studies. An early study of welder's exposure using E6XXX electrodes on mild steel showed that the fume consisted of 60 to 70% titanium dioxide, 20% silicon dioxide, and 2 to 12% manganese (7). Another early shipyard study showed that iron oxide comprised 50% of the total fume, titanium dioxide 15%, silicon dioxide 8%, and the balance was acid soluble compounds of magnesium, calcium, aluminum, manganese, chromium, copper, and sodium (8). The air concentrations of welding fume in several studies ranged from less than 5 to over 100 mg/m³ depending on the welding process, ventilation, and the degree of enclosure.

In these early studies the air samples were normally taken outside of the welding helmet. Recent studies have shown the concentration at the breathing zone inside the helmet ranges from one-third to one-sixth the outside concentrations.

Gases and vapors

Shielded metal arc welding has the potential to produce nitrogen oxides in significant concentrations; however, this is not normally a problem in open shop welding. In over 100 samples of SMA welding, the author has not identified an exposure to nitrogen dioxide in excess of 0.5 ppm under a wide range of operating conditions (9). Ozone is also fixed by the arc but again, this is not a significant contaminant in SMA welding operations. Carbon monoxide and carbon dioxide are produced from the electrode covering, but air concentrations are usually minimal.

Closed space welting

In a review of two studies of shipboard welding, the maximum concentrations of various gases noted were nitrogen dioxide (212 ppm), carbon monoxide (110 ppm), acrolein (0.08 ppm), and ozone (0.6 ppm) (10, 11). These high concentrations were observed in enclosed space welding with limited exhaust ventilation.

Radiation

The radiation generated by SMA welding covers the spectrum from the IR-C range of wavelengths to the UV-C range. To date there has been no evidence of eye damage due to infrared radiation from arc welding. The acute condition known to the welder as "arc eye," "sand in the eye," or "flash burn" is due to exposures in the UV-B range (see Figure 2.15-4). The radiation in this range is completely absorbed in the corneal epithelium of the eye and causes a severe photokeratitis. Severe pain occurs 5 to 6 hrs after exposure to the arc and the condition clears within 24 hrs. Welders experience this condition only once and then protect themselves against a recurrence by the use of a welding helmet with a proper filter, in addition to tinted safety goggles. Skin erythema or reddening may also be induced by exposure to UV-C and UV-B as shown in Figure 2.15-4.

2.15.2 Low Hydrogen Welding (Figure 2.15-1)

Low hydrogen electrodes are used with conventional arc welding systems to maintain a hydrogen-free arc environment for critical welding tasks on certain

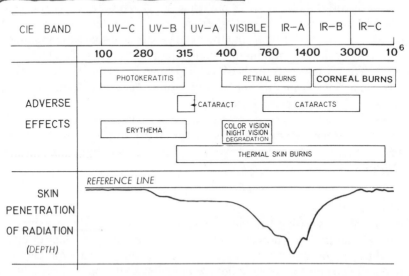

Figure 2.15-4 International Commission on Illumination (CIE) spectral bands. (From *Safety With Lasers and Other Optical Sources* by D. Sliney and M. Wolbarsht. Copyright 1980. Reproduced with permission of Plenum Publishing Co.)

steels. The electrode coating is a calcium carbonate-calcium fluoride system with various deoxidizers and alloying elements such as carbon, manganese, silicon, chromium, nickel, molybdenum, and vanadium. A large part of this coating and of all electrode coatings becomes airborne during welding. Early studies showed that 9% of the fumes from low hydrogen coating may be fluorides; of that quantity, 10 to 22% is in the form of soluble fluorides (12). Pantucek states that fluorides may comprise up to 20% of the total fume concentration depending on the coating thickness and electrode core diameter (5).

It has been suggested that the calcium fluoride is degraded in the arc to silicon hexafluoride, which then forms hydrogen fluoride in the presence of water vapor. Although hydrogen fluoride is found in the low hydrogen welding environment, it is not known if the hypothesized mechanism is correct. In addition to hydrogen fluoride, Pantucek states that sodium, potassium, and calcium fluorides are present as particulates in the fume core (5).

A comprehensive study was completed by Smith of metal fume exposure during low hydrogen welding in various industrial settings classified as confined, enclosed, and open conditions (13). In this study 75% of the samples taken in confined spaces exceed 10 mg/m³; 50% of the enclosed samples and 27% of the open samples exceeded this condition. The current TLV for fluorides of 2.5 mg/m³ was exceeded in 37% of the samples taken in enclosed spaces and in 13% of those collected in confined and open locations.

Swedish investigators have shown that low hydrogen welding on low and high alloy steels produces oxides of nickel, chromium, molybdenum, and copper in addition to the normally observed oxides of iron, silicon, and manganese. Airborne fluorides are also present in significant concentrations. In other studies, there did not appear to be a direct correlation between the percent of fluorides in the electrode coating and in the welding fume. A change in the binder from potassium to sodium silicate significantly reduces the generation rates of soluble fluorides in low hydrogen welding (14). The exposure to various gases including nitrogen dioxide and ozone during low hydrogen welding is apparently the same as that during conventional electrode welding.

Exposure to fumes from low hydrogen welding under conditions of poor ventilation may prompt complaints of nose and throat irritation and chronic nosebleeds. There has been no evidence of systemic fluorosis from this exposure. Monitoring can be accomplished by both air sampling and urinary fluoride measurements.

2.15.3 Gas Tungsten Arc (GTA) Welding (Figure 2.15-5)

Although shielded metal arc welding using coated electrodes is an effective way to weld many ferrous metals, it is not practical for welding aluminum, magnesium, and other reactive metals. The introduction of inert gas in the 1930s to blanket the arc environment and prevent the intrusion of oxygen and

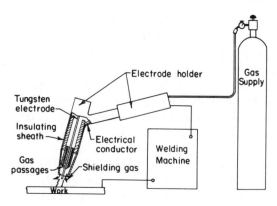

Figure 2.15-5   Gas tungsten arc (GTA) welding.

hydrogen into the weld provided a solution to this problem. In gas tungsten arc welding [also known as tungsten inert gas (TIG) welding] the arc is established between a nonconsumable tungsten electrode and the workpiece producing the heat to melt the abutting edges of the metal to be joined. Argon or helium is fed to the annular space around the electrode to maintain the inert environment. A manually fed filler rod may or may not be used. The system can be operated with either a dc or ac supply in a straight or reverse polarity mode. Straight polarity is generally used since the workpiece receives extra heat due to electron flow. The welding fume concentrations in GTA welding may be lower than manual stick welding especially if light gage material is being worked. Air concentrations of fume from manual "stick" welding will, of course, be somewhat higher if a filler rod is used. As is the case with all industrial hygiene studies of welding operations, the specific alloy and the filler wire in use must be identified. The GTA technique is used routinely on low hazard materials such as aluminum and magnesium in addition to a number of alloys including stainless steel, nickel alloys, copper-nickel, brasses, silver, bronze, and a variety of low alloy steels that may have industrial hygiene significance.

A limited number of environmental studies has been published on GTA welding. In an early study, Ferry found that the high energy GTA arc produced nitrogen dioxide concentrations at the welder's position ranging from 0.3 to 3.0 ppm with argon resulting in higher concentrations than helium (15). The GTA procedure was found to generate higher ozone concentrations than gas metal arc welding, a related procedure to be discussed in the next section, at a 0.7 m (2.3 ft) distance, but the reverse is true 1.8 m (6 ft) from the arc. In this same study the ozone concentration was at a minimum directly in the fume stream and the ozone generation rate was directly proportional to the current density. In another study, the nitrogen dioxide concentrations during GTA welding ranged from 0.048 to 2.38 mg/m³ at distances of 1 to 7 m (3.3—23 ft) from the arc with the maximum concentration noted 4 m from the arc (16).

This inert gas technique introduced a new dimension in the welder's exposure to electromagnetic radiation from the arc, with energies an order of magnitude greater than SMA welding. The energy in the UV-B and-C ranges, especially in the region of 290 nm, are the most biologically effective radiation and will produce skin erythema and photokeratitis. The energy concentrated in the wavelengths below 200 nm (UV-C) is most important in fixing oxygen as ozone. The GTA procedure does produce a rich, broad spectral distribution with important energies in these wavelengths. Kranz states that the arc spectrum is dependent on the main metal component of the electrode and that the alloying metals and electrode cover constituents do not play a significant role in defining the spectral distribution (17). Dahlberg's spectral data on steel welding with a variety of coated electrodes and gas shielded electrodes apparently supports this statement (18). In inert gas welding of aluminum, the spectrum maximum is stated to be aluminum lines. Early studies by Silverman showed that the ozone concentration was higher when welding on aluminum than on steel due to the UV reflectance of aluminum and that argon produced higher concentrations of ozone than of helium due to its stronger spectral emission. The spectral energy is shown to be dependent on current density (19).

From the preceding discussion it is obvious that air sampling must be conducted for both ozone and nitrogen dioxide when evaluating exposures from GTA welding.

When the GTA process was introduced, it was found that a tungsten alloy containing 2% thoria was desirable for the nonconsumable electrode. This prompted concern with regard to the potential problem from the airborne levels of radioactive thoria. Studies showed that under normal current the electrode was consumed at a rate of 0.1 to 0.3 mg/min (20). The airborne levels of thoria at the breathing zone were not significant although ventilation control was recommended when dressing the thoriated-tungsten electrode.

2.15.4 Gas Metal Arc (GMA) Welding (Figure 2.15-6)

In the 1940s a consumable wire electrode was developed to replace the nonconsumable tungsten electrode used in the GTA system. Originally developed to weld thick, thermally conductive plate, this GMA process [also known as manual inert gas (MIG) welding] now has widespread application on aluminum, copper, magnesium, nickel alloys, titanium, as well as steel alloys.

In this system the welding torch has a center consumable wire that maintains the arc as it melts into the weld puddle. Around this electrode is an annular passage for the flow of helium, argon, carbon dioxide, nitrogen, or a blend of these gases. The wire usually has the same or similar composition to the base metal with a flash coating of copper to insure electrical contact in the gun and to prevent rusting. The filler wire usually contains deoxidizers or scavengers. Manganese, silicon, and aluminum are used as deoxidizers for steel

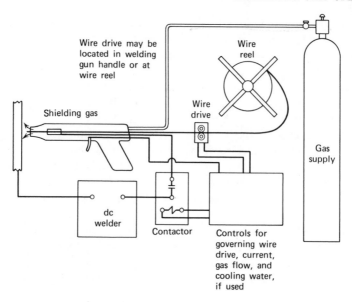

Figure 2.15-6 Gas metal arc (GMA) welding.

filler wires; titanium and silicon, for nickel alloy wire; and titanium, silicon, and phosphorus, for copper alloy wires.

An improvement in GMA welding is the use of a flux-cored consumable electrode. The electrode is a hollow wire with the core filled with various deoxidizers, fluxing agents, and metal powders. The arc may be shielded with carbon dioxide or the inert gas may be generated by the flux core.

Alpaugh studied the differences between GMA welding and shielded metal arc welding and found that the iron oxide fume concentrations with SMA welding were generally higher than GMA welding (21). The nitrogen dioxide concentrations were of the same order of magnitude; however, ozone concentrations were higher with the GMA technique. Two separate investigations indicate that GMA welding produces higher concentrations of ozone than GTA welding (22, 23). The ozone generation rate increases with an increase in current density but plateaus rapidly. The arc length and the inert gas flow rate do not have a significant impact on the ozone generation rate. Nitrogen dioxide concentrations during GMA welding have not been thoroughly investigated although one investigator indicated that if ozone concentrations are high, nitrogen dioxide concentrations will be low and vice versa (24).

Carbon dioxide is widely used on GMA welding due to its attractive price; argon and helium cost approximately 15 times as much as carbon dioxide. The carbon dioxide process is similar to other inert gas arc welding shielding techniques and one encounters the usual problem of metal fume, ozone, oxides of nitrogen, decomposition of chlorinated hydrocarbons solvents, and UV radiation. In addition, the carbon dioxide gas is reduced to form significant concentrations of carbon monoxide. The generation rate of carbon monoxide depends on current density, gas flow rate, and the base metal being welded. Although the concentrations of carbon monoxide may exceed 100 ppm in the fume cone, the concentration drops off rapidly with distance and with reasonable ventilation, hazardous concentrations should not exist at the breathing zone.

The intensity of the radiation emitted from the arc is, as in the case of GTA welding, an order of magnitude greater than that noted with shielded metal arc welding. The impact of such a rich radiation source in the UV-B and UV-C wavelengths has been covered in the GTA discussion. Both procedures present an additional problem first described by Ferry (15). Trichloroethylene and other chlorinated hydrocarbon vapors are decomposed by UV emission from the GMA arc forming chlorine, hydrogen chloride, and phosgene. The solvent is degraded in the UV field and not directly in the arc. The condition was noted only with inert gas shielded techniques, especially when argon was used as the shielding gas. Studies by Dahlberg showed that hazardous concentrations of phosgene could occur with trichloroethylene and 1,1,1-trichloroethane even though the solvent vapor concentrations were below the appropriate TLV for the solvent (25). The GMA technique will produce higher concentrations of phosgene than GTA welding under comparable operating conditions. Dahlberg also identified dichloroacetyl chloride as a principle product of decomposition that could act as a warning agent due to its lacrymatory action. In a follow-up study, perchloroethylene was shown to be less stable than other chlorinated hydrocarbons. When this solvent was degraded, phosgene was formed rapidly, and dichloroacetyl chloride was also produced in the UV field (26). In the author's opinion perchloroethylene represents a greater hazard than trichloroethylene or 1,1,1-trichloroethane from UV transformation.

$COCl_2$

In summation, perchloroethylene, trichloroethylene, and 1,1,1-trichloroethane vapors in the UV field of high energy arcs, such as those produced by GTA and GMA welding, may generate hazardous concentrations of phosgene and should be evaluated by air sampling. Other chlorinated solvents may present a problem depending on the operating conditions; therefore, diagnostic air sampling should be performed. The principle effort should be in the control of the vapors to insure that they do not appear in the UV field. Delivery of parts directly from degreasing to the welding area has, in the experience of the author, presented major problems due to pullout and trapping of solvent in the geometry of the part.

2.15.5 Submerged Arc Welding (Figure 2.15-7)

In this process, shielding of the arc from the atmosphere is accomplished by covering the weld with a granular, fusible flux. The filler metal is a bare wire electrode; in addition, one may use a supplemental filler wire to feed into the arc. The granular flux is fed onto the metal ahead of the arc path, sintering to form a molten slag cover over the weld metal. The flux shields the arc, adds

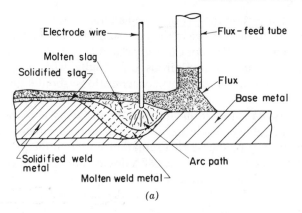

(a)

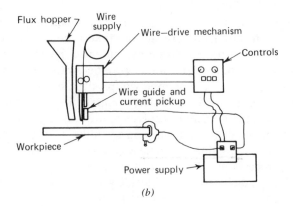

(b)

Figure 2.15-7 Submerged arc welding. (a) Weld cross section. (b) Welding equipment.

alloy metal, stabilizes the arc, and defines the weld-bead shape. The technique is used in either a semi- or fully-automatic mode for welding thick sections on plain carbon and low alloy steel.

As expected, the metal fume concentrations during submerged arc welding are lower than those for either SMA and GMA welding due to the blanketing action of the flux. The arc is maintained beneath the flux without sparks, smoke, or flash. One author states that this method produces as little as one-eighth the fume of other arc welding procedures (27). An analysis of the fume from submerged arc welding shows significant concentrations of silicon dioxide, iron oxide, fluorides, and manganese (28).

The principle hazard in this procedure is hydrogen fluoride and soluble fluoride particulates in the air released from the welding flux. Since the flux eliminates the direct arc radiation and suppresses the generation of metal fume, submerged arc welding has not received significant industrial hygiene review. In most cases, the process is not provided with local exhaust ventilation but relies on dilution ventilation.

2.15.6   Plasma Arc Welding (Figure 2.15-8)

In the plasma arc process the welding head is designed to provide a flow of a gas such as argon through an orifice under a high voltage gradient resulting in a highly ionized gas stream. A complex interaction of mechanical and electromagnetic forces produces arc temperatures greater than 33,400°C (60,000°F). In addition to welding, this technique is widely used for cutting and metallizing. The hazards of the process include noise, ozone, nitrogen dioxide, and metal fume exposures.

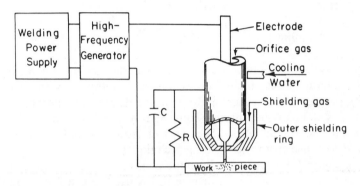

Figure 2.15-8   Plasma arc welding.

The health hazards presented by plasma arc welding are similar to those presented by GTA welding but they introduce some new problems. The UV spectrum from plasma arc welding is much more intense than other inert gas arc welding systems, resulting in major skin and eye exposures requiring special clothing and eye protection. For example, plasma normally requires the use of a No. 14 shade and full chrome leather clothing, while light manual SMA welding requires only a No. 10 shade. The rich UV-B spectrum (280-315 nm) results in high ozone generation rates and the rich arc is effective in fixing nitrogen oxides. Fannick and Corn have shown that the concentrations of ozone and nitrogen dioxide can easily exceed the TLVs unless local exhaust ventilation is available (29).

In a review of seven studies, the ozone concentration during plasma welding ranged from not detectable to 8.3 ppm during the cutting of steel without ventilation. The nitrogen dioxide concentrations reached a maximum of 9.6 ppm under these conditions. The noise level at the operator's position may be in the range of 110–120 dBA. This noise is principally aerodynamic in origin from the plasma stream and is therefore difficult to control. Many operations require an enclosed operating booth for noise control with a downdraft exhausted welding table for removal of air contaminants.

2.15.7   Resistance Welding (Figure 2.15-9)

In resistance welding, an electric current is passed through workpieces held together under pressure. There is a localized heating at the contact surfaces due to the contact resistance, and the metal coalesces. Various welding methods including seam, spot, projection, and flash welding are based on this technique. No flux or filler metal is added to this process.

Resistance welding is used widely for the assembly of light sheet metal fabrications. The hazards are minimal; indeed, the only complaint arising from this operation is due to welding parts with residual surface oil, which degrades during the welding, forming aldehydes that prompt worker complaints from olfactory, respiratory, and eye irritation.

2.15.8   Gas Welding (Figure 2.15-10)

In this process the heat of fusion is obtained from the combustion of oxygen and one of several gases including acetylene, methylacetylene-propadiene (MAPP), propane, butane, and hydrogen. The flame melts the workpiece and a filler rod is manually fed into the joint. Gas welding is used widely for light sheet metal and repair work. The hazards from gas welding are minimal compared to those from arc welding techniques.

The uncoated filler rod is usually of the same composition as the metal being welded except on iron where a bronze rod is used. Paste flux is applied by dipping the rod into the flux. Fluxes are used on cast iron, some steel alloys, and nonferrous work to remove oxides or assist in fusion. Borax-based fluxes are used widely on nonferrous work while chlorine, fluorine, and bromine compounds of lithium, potassium, sodium, and magnesium are used on gas welding of aluminum and magnesium.

The metal fume originates from the base metal, filler metal, and the flux. The fume concentration encountered in field welding operations depends principally on the degree of enclosure in the work area and the quality of

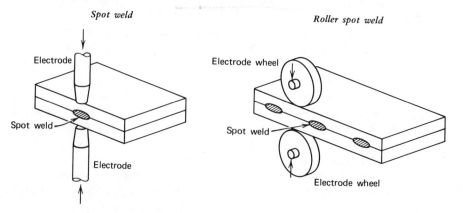

Figure 2.15-9   Resistance welding.

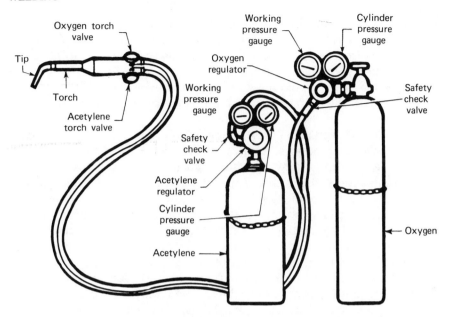

Figure 2.15-10   Gas welding. (Courtesy of Hobart Brothers Co.)

ventilation. Since gas or torch welding is conducted at temperatures lower than those for arc welding processes, one seldom encounters excessive concentrations of metal fumes except when using lead, zinc, and cadmium, which have significant vapor pressures at relatively low temperatures.

The principal hazard in gas welding in confined spaces is due to the formation of nitrogen dioxide. Dana states the oxides of nitrogen are formed from the nitrogen present in the oxygen supply; another investigator demonstrated that the concentrations of nitrogen dioxide were 200 ppm while using 98% pure oxygen and 180 ppm while using 99.5% pure oxygen (30). Higher concentrations occurred when the torch was burning without active welding. Striekevskia found concentrations of nitrogen dioxide of 280 mg/m³ in a space without ventilation and of 12 mg/m³ in a space with some ventilation (31). This latter investigator cautions that phosphine may be present as a contaminant in acetylene and that carbon monoxide may be generated during heating of cold metal with a gas burner.

The radiation from gas welding is quite different from arc welding. The principal emissions are in the visible and IR-A,-B,-C wavelengths and require the use of light tinted goggles for work. Ultraviolet radiation from gas welding is negligible.

2.15.9   Scarfing and Cutting

Several techniques have been developed to cut slab steel, remove gates and riser systems on castings, gouge out defective metal in castings, and remove surface scale on billets. The Arcair process utilizes a copper-clad, boron-graphite electrode in a manual electrode holder with an integral compressed air supply to blow away the molten metal from the puddle as it melts. Electrode diameters vary from 3.2 to 25.4 mm (1/8 to 1 in.) and operate at 50 to 2200 A to remove up to 0.9 kg (2.0 lb) of metal per minute. A later process, powder burning, is based on an oxygen-acetylene torch with the addition of iron powder to create a high temperature flame that will cut through very thick steel metal sections.

The principal hazard from these operations is the exposure to metal fume. In a study of these processes, a series of breathing zone samples inside the helmet on a powder burning operator averaged 31.1 mg/m³ without local exhaust ventilation and less than 7 mg/m³ when control was achieved with sidedraft local exhaust ventilation (32). On an Arcair operation, the average of the inside helmet samples was 21.5 mg/m³ but dropped to 3.3 mg/m³ when local exhaust was operating. The authors of the study state that the worker would be exposed to metal fumes of manganese, nickel, copper, and chromium, in addition to carbon monoxide, oxides of nitrogen, and ozone.

In a study conducted in the United Kingdom, Sanderson reviewed arc gouging operations in open, semi-enclosed, and enclosed settings and found that in ventilated operations the principal hazards were iron oxide fume from the workpiece and copper fume from the cladding on the electrode (33). The air concentrations of copper were 14 times the TLV. In enclosed operations, lead, copper, and iron oxide fume concentrations were high and ozone and carbon monoxide concentrations exceeded their respective TLVs.

The effectiveness of local exhaust booths on Arcair operations in a foundry were evaluated in a recent NIOSH-sponsored study and a summary of the air concentration data is shown in Table 2.15-3 (34). Several conclusions were drawn from this valuable study. The authors found that there was no significant difference between lapel and in-helmet samples and that individual work practice had a significant impact on the exposure to metal fume. In conventional welding, air concentrations inside the helmet may be as low as one-fifth to one-tenth the ambient air concentrations. Despite rather good ventilation control, TLVs for total welding fume or iron oxide fume were exceeded routinely. If time weighted concentrations were calculated based on full shift operations, only two out of four of the workers exceeded the TLV for total and iron oxide fume. The air jet caused violent dispersal of the metal fume, which made it difficult to achieve efficient collection by the hood. The authors concluded that control could be established with booth ventilation but that individual work practice is an important consideration.

2.15-10 Control of Exposure

Metal fume exposure

In the discussions of the individual processes it was assumed that the workpiece was mild steel and, therefore, that the predominant metal fume contaminant was iron oxide fume. One must also consider the fume exposures that occur while welding steel alloys, nonferrous metal, and copper alloys. Metal coatings and the welding electrodes also contribute to the metal fume exposure. Table 2.15-4 shows the wide range of metal fumes encountered in welding. Recent interest in the possible carcinogenicity of chromium and nickel fume has prompted special attention to these exposures during welding. Lead has been used as an alloy in steels to improve its machinability, and welding on such material requires rigorous control. This is also the case with manganese used in steel alloys to improve metallurgical properties. Beryllium, probably the most toxic alloying metal, is added to copper and warrants close control during welding and brazing operations.

Welding or cutting on workpieces that have metallic coatings may be especially hazardous. Lead-based paints have been used commonly on marine and structural members. Welding on these surfaces during repair and shipbreaking generates high concentrations of lead fume. In one shipyard study, lead concentrations ranged from 0.3 to 200 mg/m³ (35). Another shipyard study showed air concentrations of lead of 0.1 to 0.65 mg/m³ (36). In the author's experience, cutting and welding on structural steel covered with lead paint resulted in concentrations exceeding 1.0 mg/m³ in well-ventilated conditions out of doors.

Steel is galvanized by dipping in a hot zinc dip. Pesques found that air concentrations of zinc during the welding of galvanized steel and steel painted with zinc silicate ranged from 3 to 12 times the TLV under conditions of poor ventilation (37). The author noted that the concentrations were lower with oxygen-acetylene torch work than with arc cutting methods. When good ventilation was established, the TLV for zinc was seldom exceeded.

Venable states that the hazard from burning and welding of pipe coated with a zinc-rich silicate can be minimized in two ways (38). Since pipes are joined end to end, the principle recommendation is to mask the pipe ends with tape before painting. If it is necessary to weld pipe after it is painted, the first step before welding is to remove the paint by hand filing, power brush or grinding, scratching, or abrasive blasting. If the material cannot be removed, suitable respiratory protection is mandated; in many cases air-supplied respirators may be required.

Conventional arc welding (SMA) on ferrous metals in open areas can usually be performed safely with dilution ventilation; however, welding in enclosed spaces will always require local exhaust. Local exhaust ventilation is also needed when using GTA or GMA techniques on stainless steel, high alloy steels, nickel alloys, copper alloys, or when the base metal is coated with toxic metal. Guidelines for ventilation control by dilution and conventional local exhaust are available (39). A high velocity-low air volume exhaust system has been evaluated for several welding applications with mixed reviews (40).

The most difficult control problem is welding in confined spaces such as on shipboard. In the 1940s Viles and Silverman developed a portable exhaust system unit for shipboard use with an integral particulate filter for removal of metal fume with direct recirculation. A similar system with both a particulate filter and an adsorption system for the removal of gases has been introduced recently in Denmark and may have application on operations when one cannot utilize conventional local exhaust systems.

Ozone and oxides of nitrogen

As noted in Table 2.15-1, these contaminants are generated by diverse welding procedures; however, the use of inert gas shielding operations in confined locations or with poor exhaust ventilation presents the major problem.

Since ozone is fixed by UV-C (most efficiently by wavelengths less than 200 nm), shielding of the arc may be effective. As mentioned earlier, in many cases mechanical shielding of the arc is not possible and the only control is exhaust ventilation. Hazardous concentrations of nitrogen dioxide may be generated in enclosed spaces in short periods of time and, therefore, require effective exhaust ventilation. Techniques for improved ventilation in enclosed refinery vessels have been proposed by Brief (41).

Radiant energy

Eye protection from exposure to UV-B and UV-C wavelengths is obtained with filter glasses in the welding helmet of the correct shade, as recommended by the American Welding Society. The shade choice may be as low as 8–10 for light manual electrode welding or as high as 14 for plasma welding. Eye protection must also be afforded other workers in the area. To minimize the hazard to nonwelders in the area, flash screens or barriers should be installed. New plastic "see-through" screens are effective in minimizing UV radiation. Plano goggles or lightly tinted safety glasses may be adequate if one is some distance from the operation. If one is 9 to 12 m (30 to 40 ft) away, eye protection is probably not needed for conventional welding; however, high current density GTA or GMA welding may require that the individual be 30.5 m (100 ft) away before direct viewing is possible without eye injury. On gas welding and cutting operations, IR wavelengths must be attenuated by proper eye protection for worker health and comfort.

The attenuation afforded by tinted lens is available in NBS reports and the quality of welding lens has been evaluated by NIOSH (42). Glass safety goggles will provide some protection from UV; however, plastic safety glasses may not.

The ultraviolet radiation from inert gas shielded operations will cause skin erythema or reddening; and therefore the welder must adequately protect his or her face, neck, and arms. Heavy chrome leather vest armlets and gloves must be used with such high energy arc operations.

Table 2.15-3 *Air Contaminant Levels of Metals and Total Dust (TD) in mg/m³*

	Fe	Cu	Cr	Mn	Zn	Ni	TD
Geometric mean	3.32	0.030	0.028	0.093	0.0052	0.00214	9.62
Standard deviation	2.78	2.79	4.51	2.70	2.80	4.18	2.72

Source.   Reference 34.

Table 2.15-4 *Contaminants from Welding Operations*

Contaminant	Source
Metal fumes	
Iron	Parent iron or steel metal, electrode
Chromium	Stainless steel, electrode, plating, chromate primed metal
Nickel	Stainless steel, nickel-clad steel
Zinc	Galvanized or zinc primed steel
Copper	Coating on filler wire, sheaths on air-carbon arc gouging electrodes, nonferrous alloys
Vanadium, manganese, and molybdenum	Welding rod, alloys in steel
Tin	Tin coated steel
Cadmium	Plating
Lead	Lead paint, electrode coating
Fluorides	Flux on electrodes
Gases and vapors	
Carbon monoxide	CO_2 shielded, GMA, carbon arc gouging, oxy-gas
Ozone	GTA, GMA, carbon arc gouging
Nitrogen dioxide	GTA, GMA, all flame processes

Decomposition of Chlorinated Hydrocarbon Solvents

The decomposition of these solvents occurs in the UV field around the arc and not in the arc itself. The most positive control is to prevent the solvent vapors from entering the welding area in detectable concentrations. Merely maintaining the concentration of solvent below the TLV is not, in itself, satisfactory. If vapors cannot be excluded from the workplace, the UV field should be reduced to a minimum by shielding the arc. Pyrex® glass is an effective shield that permits the welder to view his or her work. Rigorous shielding of the arc is frequently possible at fixed station work locations, but it may not be feasible in field welding operations.

REFERENCES

1 American Welding Society, *Fundamentals of Welding and Welding Processes—Arc and Gas Welding and Cutting, Brazing and Soldering,* Vols. 1 and 2 of *Welding Handbook,* 7th ed., AWS, Miami, FL, 1978.

2 American Welding Society, *The Welding Environment: A Research Report on Fumes and Gases Generated During Welding Operations,* AWS, Miami, FL, 1973.

3 American Welding Society, *Fumes and Gases in the Welding Environment,* AWS, Miami, FL, 1979.

4 American Welding Society, "Filler Metal Procurement Guides," Publication No. 5.01-78, AWS, Miami, FL, 1978.

5 M. Pantucek, *Am. Ind. Hyg. Assoc. J.,* **32,** 687 (1971).

6 H. Buck and A. Dessler, International Institute of Welding Document III-311-67, Paris, France, 1967.

7   B. Tebbens and P. Drinker, *J. Ind. Hyg. Tox.*, **23**, 322 (1941).

8   M. Callen, *J. Ind. Hyg. Tox.*, **29**, 113 (1947).

9   Wm. A. Burgess, personal observations.

10 W. Kierst, J. Vselis, M. Graczyk, and A. Krynick, *Bull. Inst. Mar. Med. Gdansk,* **15,** 149 (1964).

11 J. Steel, *Ann. Occup. Hyg.,* **11,** 115 (1978).

12 J. Ferry, *Am. Ind. Hyg. Assoc., Q.,* **14,** 173 (1953).

13 L. Smith, *Ann. Occup. Hyg.,* **10,** 113 (1967).

14 K. Dawada and K. Twano, International Institute of Welding Document VIII-197-64, Paris, France, 1964.

15 J. Ferry and G. Ginter, *Welding J.,* **32,** 396 (1953).

16 O. Opris and N. Ionsseu, International Institute of Welding Document VIII-342-68, Paris, France, 1968.

17 E. Kranz, *Schwisstechnik (Berlin),* **15,** 446 (1965).

18 J. Dahlberg, *Ann. Occup. Hyg.,* **14,** 259 (1971).

19 L. Silverman and H. Gilbert, *Welding J.,* **33,** 218 (1954).

20 A. Breslin and W. Harris, *Am. Ind. Hyg. Q.,* **13,** 191 (1952).

21 E. Alpaugh, K. Phillips, and H. Pulsifer, *Am. Ind. Hyg. Assoc. J.,* **29,** 551 (1968).

22 R. Frant, *Ann. Occup. Hyg.,* **6,** 113 (1963).

23 F. Lanau, *Ann. Occup. Hyg.,* **10,** 175 (1967).

24 E. I. Vorontsova, T. S. Karacharov, and K. Voshcanov *Welding Prod. (USSR),* **7,** 59, (1961).

25 J. Dahlberg, International Institute of Welding Document VIII 341-68, Paris, France, 1968.

26 H. Anderson, J. Dahlberg, and R. Wettstrom, *Ann. Occup. Hyg.,* **18,** 129 (1975).

27 S. Mosendz, L. Zaitskaya, and A. Pines, *Auto. Weld.,* **21,** 74 (1968).

28 S. Byczkowski, A Bondanowica, W. Kopczyniski, and W. Senczuk, International Institute of Welding Document VIII-196-64, 1964.

29 N. Fannick and M. Corn, *Am. Ind. Hyg. Assoc. J.,* **30,** 226 (1969).

30 J. Dana, *Using Nouv.,* **4,** 15 (1948).

31 I. Strizkerskiy, *Weld. Prod.,* **7,** 40 (1961).

32 F. Sentz and A. Rakow, *Am. Ind. Hyg. Assoc. J.,* **30,** 143 (1969).

33 J. Sanderson, *Ann. Occup. Hyg.,* **11,** 123 (1968).

34 Anonymous, "An Evaluation of Occupational Health Hazard Control Technology for the Foundry Industry," U.S. Department of Health, Education and Welfare, Publication No. (NIOSH) 79-114, Cincinnati, OH, 1979.

35 J. Steel, *Ann. Occup. Hyg.,* **11,** 115 (1968).

36 A. Jones, International Institute of Welding Document VIII-171-63, Paris, France, 1963.

37 W. Pesques, *Am. Ind. Hyg. Assoc. J.,* **21,** (3), 252–255, (1960).

38 F. Venable, *Esso Med. Bull.,* **39,** 129 (1979).

39 Committee on Industrial Ventilation, American Conference of Governmental Industrial Hygienists, *Industrial Ventilation: A Manual of Recommended Practice,* 16th ed., ACGIH, Lansing, MI, 1980.

40 H. VanWagenen, "Assessment of Selected Control Technology Techniques for Welding Fumes," Department of Health, Education and Welfare, Publication No. (NIOSH) 79-125, Cincinnati, OH, 1979.

41 R. S. Brief, L. Raymond, W. Meyer, and J. Yoder, *Air Cond. Heat Vent.,* **58,** 73 (1961).

42 D. Campbell, "Report on Tests of Welding Filter Plates," Department of Health, Education and Welfare, Publication No. (NIOSH) 76-198, Cincinnati, OH, 1976.

PRODUCTION FACILITIES

3

3.1 ABRASIVES

In Chapter 2, the use of abrasives for heavy-duty cleaning of metal surfaces, machining, grinding, buffing, and polishing is noted. The abrasives commonly in use in modern industry are shown in Table 3.1-1. The natural occurring abrasives have been replaced to a major degree by the synthetic abrasives, aluminum oxide (Al_2O_3), and silicon carbide (SiC). These two abrasives are used in loose granular form for abrasive blasting, in bonded grinding wheels for dimensional and nondimensional work, and in coated products such as abrasive paper and cloth. The specialty abrasives, industrial diamonds or cubic boron nitride, are coated in a thin layer on solid metal wheels for grinding hard metals and other abrasives.

Table 3.1-1 *Natural and Synthetic Abrasives*

Abrasive	Formula	Manufacturer Identification
Natural		
Quartz	SiO_2	Silica sand, sandstone, flint, tripoli
Natural aluminum oxide	Al_2O_3	Corundum, emery
Iron oxide	Fe_2O_3	Rouge, crocus
Synthetic		
Aluminum oxide	Al_2O_3	Alundum, aloxite
Silicon carbide	SiC	Carborundum, crystolon
Diamond	Pure C	Man-made
Cubic boron nitride	BN	Barozon

Aluminum oxide is made by calcining bauxite (aluminum oxide ore) to remove water and then firing the bauxite in an electric furnace with coke and iron chips that purify the batch (1). After the material is fired at a temperature of 2040°C (3700°F), the fused mass or pig is crushed, granulated, and sized for a given abrasive product. The product has a range of colors from the purest white form to yellow, green, and blue depending on the metal compounds added to provide cutting properties. The potential hazards in the manufacture of aluminum oxide include exposures to carbon monoxide, particulates, heat, and noise.

Silicon carbide is also made in a batch operation by the heating of pure quartz, sodium chloride, sawdust, and coke. The material is packed over a graphite resistance electrode and heated to 2000°C (3630°F) to form silicon carbide.

$$SiO_2 + 3C \rightarrow SiC + 2CO$$

The sawdust burns providing porosity to vent the pig while the salt assists in the removal of impurities. When the relatively pure silicon carbide is formed at the center of the mass, it is cooled, the mass is crushed, the abrasive in the core of the pig is removed for processing, and the skin of the pig is recycled (1). The color of the abrasive varies from green to black depending on the impurities present. The release of massive quantities of carbon monoxide presents the major hazard in the manufacture of silicon carbide though exposure to crystalline silica and heat stress are also significant.

Diamond and boron nitride are extremely hard abrasives that have up to 100 times the life of aluminum oxide and silicon carbide. The diamond abrasive material is formed by heating graphite and a catalyst such as nickel to 1500–2000°C (2730–3630°F) at 6.5–9 MPa. Cubic boron nitride is formed from hexagonal boron nitride with a lithium compound as a catalyst at 2000°C (3630°F) under 8 MPa pressure. In the open literature, no data on the potential health hazards from these operations are available.

Silicon carbide and aluminum oxide may be formed into abrasive grinding wheels by various bonding techniques, the most common of which are vitrified and resinoid bonds. To manufacture a vitrified bonded wheel, the abrasive is mixed with clay or feldspar. This mix is then poured into a cavity, pressed to shape, and fired at 430–650°C (800–1200°F) to form a structurally strong wheel with a glass or ceramic bonding matrix. The batch material for resinoid wheels shown in Table 3.1-2 includes the abrasive and a resin such as phenol-formaldehyde. The blended material is pressed to shape and heated at 160–200°C (320–390°F). The cured rough wheels are then trued or dressed to shape and a center bushing is added using lead or babitt metal.

The coated abrasive products are manufactured by applying the abrasive grains to paper or cloth covered with hide glue or phenolic resin. Polishing wheels are flexible discs of laminated cloth or leather and the abrasive is added to the perimeter surface of the wheel using a resin glue to bond the abrasive. The potential exposures include particulates and the volatile components from the adhesive system.

Table 3.1-2 *Typical Materials
for Resinoid Wheel*

Solids
Olivine sand
Synthetic cryolite
Quicklime
Litharge
Activated charcoal
Abrasive granules
Resins
   Phenolic resin
   Thoric acid
   Potassium sulfate
   Iron pyrite
Liquids
Furfural
Creosote
Liquid resins
   Polymerized phenol-formaldehyde
   Free phenols
   Sodium hydroxide catalyst

Source.   Reference 2.

During the manufacture of abrasive products there may be exposure to clays and fillers, such as feldspars in the manufacture of vitrified bonded wheels and shellac, rubber, and various resins systems in the production of resinoid and other bonded wheels. Strict ventilation control of dust should be maintained during crushing, grinding, and screening of the bulk abrasive as well as in the edging, facing, and shaving of grinding wheels (Table 3.1-3). A dermatitis problem may exist depending on the resin systems and adhesives in use.

Table 3.1-3 *Ventilation for Abrasive Wheel Manufacture*

	Ventilation	
Operation	Type of Hood	Capture Velocity at Hood Face (fpm)
Grading screen	Enclosure booth	50
Barrels	Close canopy	400
Grinding wheel dressing	Enclosure booth	400

Source.   Reference 3.

REFERENCES

1   L. Coes, Jr., *Abrasives,* Springer-Verlag, Bonn, 1971
2   Health Hazard Evaluation Determination Report No. 73-18-171, U.S. Department of Health, Education and Welfare, NIOSH, Cincinnati, OH, 1975
3   Committee on Industrial Ventilation, American Conference of Governmental Industrial Hygienists, *Industrial Ventilation: A Manual of Recommended Practice,* 14th ed., ACGIH, Lansing, MI, 1976

3.2   ACIDS

3.2.1   Hydrochloric Acid

Hydrogen chloride may be prepared by reacting sulfuric acid with sodium chloride to form sodium bisulfite, which then reacts with sodium chloride to form hydrogen chloride, and sodium sulfate. The acid is made by absorbing hydrogen chloride gas in water in absorption towers.

The bulk of the hydrochloric acid produced in the United States is formed as a by-product in the chlorination of organic compounds. In the presence of a catalyst, benzene, chlorine, and hydrogen are reacted to form chlorobenzene; hydrogen chloride is a reaction product. Benzene and chlorobenzene are first recovered, and the hydrogen chloride in the final stream is removed in a falling film absorption tower.

The principal hazard in these processes is exposure to either the leakage of gas and vapor from the system or to the tail gas from the scrubber. Present absorption tower design is adequate to minimize exposure of operating personnel and the neighborhood. The safety and handling procedures for hydrochloric acid have been published and include operating practice, protective clothing, eye protection, and respiratory protection (1).

3.2.2   Nitric Acid

The principal method for manufacturing nitric acid is the high pressure ammonia-oxidation process (Figure 3.2-1) in which air and ammonia (9.5–11%) are passed over a heated platinum-rhodium catalyst with the resultant oxidation of ammonia to yield nitric oxide (NO). In this process the concentration of ammonia must be kept below the lower limit of flammability (15.5% in air, 14.8% in oxygen) to avoid an explosion. The nitric oxide is oxidized to nitrogen dioxide (NO_2) in the reaction and is absorbed by water in a bubble cap plate column to yield 50–70% nitric acid (HNO_3).

Nitric acid recovery plants may represent a significant exposure to oxides of nitrogen due to liquid or gas leaks and during on-stream sampling. The emissions from exhaust stacks or nitrogen dioxide scrubbers may also result in hazardous conditions at ground level since the concentration of total oxides of nitrogen in the tail gas stream may be as high as 0.3%. Air cleaning of the

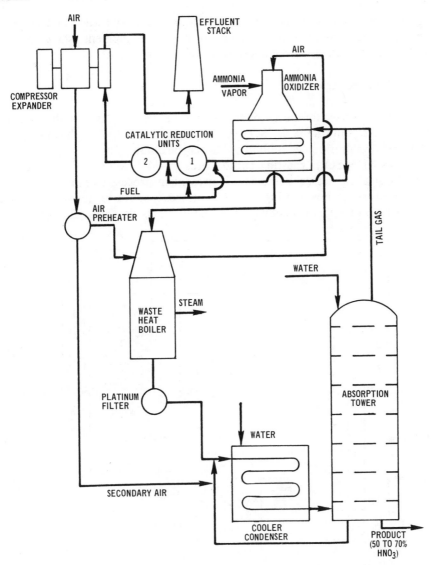

Figure 3.2-1 Flow diagram of typical nitric acid plant using pressure process. (From Reference 6)

exhaust stream is based on the reduction of the oxides or absorption in alkaline liquors. The safety hazards in the handling of this acid are well documented and are, in general, those assigned to strong oxidizers.

Accidental exposure to ammonia also may occur because of escape of gas from storage tanks, gauge glasses, valves, and process lines. In addition to the health hazard from ammonia, a potential fire and explosion hazard may exist.

The serious hazards from nitrogen dioxide and ammonia necessitate a rigorous respiratory protection program for emergency escape and reentry. The usual requirements for personal protective equipment, showers, and well-defined work practices apply to nitric acid plants (2).

3.2.3 Sulfuric Acid

The contact process for the manufacture of sulfuric acid is based on the catalytic conversion of sulfur dioxide (SO_2) to sulfur trioxide (SO_3) with the absorption of the SO_3 in sulfuric acid. Sulfur dioxide may be obtained by burning elemental sulfur (8–11% SO_2), roasting sulfide ores (7–14% SO_2), or from various metallurgical processes where SO_2 is present in process streams.

The process, shown in Figure 3.2-2, utilizes a converter tower containing several beds of a pentavalent vanadium catalyst in pellet form. The reaction in

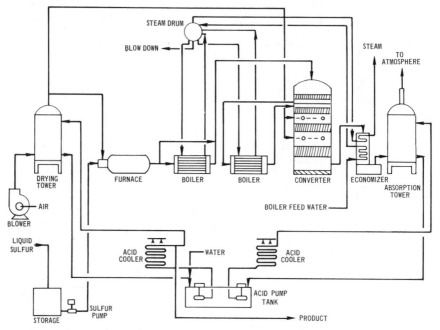

Figure 3.2-2 Basic flow diagram of contact-process sulfuric acid plant burning elemental sulfur. (From Reference 6)

the converter ($2SO_2 + O_2 \rightarrow 2SO_3$) is conducted at 400-600°C (750-1110°F). This stream is directed to the sulfuric acid absorption tower where tail gas concentrations of SO_2 may be as high as 2000 ppm. In a modification of this process, double catalysis or interpass absorption, a stream is taken off prior to the last converter stage and conveyed directly to the absorber. This modification reduces the SO_2 concentrations in the tail gas stream to 100–300 ppm.

Significant SO_2 and SO_3 air concentrations at sulfuric acid plants are attributable to fugitive leaks and off-gas emissions; therefore, both emergency escape and reentry respiratory protective equipment must be available.

Effective control may be achieved by maintenance of the closed process and the use of material handling techniques to minimize employee contact. Retaining walls should be available in case holding tanks and drums are ruptured. Acid should not be stored near reducing agents because of the fire and explosion hazard. Impervious protective clothing, eye protection, and emergency showers should be available at the workplace (3, 4, 5).

The emission of ammonium vanadate or vanadium pentoxide catalyst from the contact process may present a hazard to workers (6).

REFERENCES

1 Chemical Manufacturers Association, "Hydrochloric Acid, Aqueous, and Hydrogen Chloride, Anhydrous," Chemical Safety Data Sheet No. 5D-39, CMA, Washington, D.C., 1951

2 Chemical Manufacturers Association, "Properties and Essential Information for Safe Handling and Use of Nitric Acid," Chemical Safety Data Sheet No. SD-5, CMA, Washington, D.C., 1961.

3 O. T. Fasullo, *Sulfuric Acid, Use and Handling*, McGraw-Hill, New York, 1965.

4 "Criteria for a Recommended Standard—Occupational Exposure to Sulfuric Acid," U.S. Department of Health, Education and Welfare, Publication NO. (NIOSH) 74-128, 1974.

5 Chemical Manufacturers Association, "Properties and Essential for Safe Handling and Use of Sulfuric Acid," Chemical Safety Data Sheet No. SD-20, CMA, Washington, D.C., 1963.

6 "Compilation of Air Pollutant Emission Factors," 2nd ed. U.S. Environmental Protection Agency, Publication No. AP-42, Research Triangle Park, 1973.

3.3 ALUMINUM

Aluminum is produced by the electrolytic reduction of alumina (Al_2O_3) which is obtained from bauxite, an ore produced primarily in South America and Australia (1). Bauxite is an earthy-white to reddish mineral composed of Al_2O_3, Fe_2O_3, TiO_2 and SiO_2. The highest grade ore contains 55% Al_2O_3 and has a maximum of 8% SiO_2.

The process flow in the manufacture of aluminum is shown in Figure 3.3-1 (2). The bauxite is first washed to remove clay and other waste and then dried prior to grinding to a defined particle size in a ball mill. It is refined by dissolving the bauxite in a hot caustic solution, precipitating the purified aluminum hydroxide, and calcining the hydrate to form Al_2O_3.

The handling of the ore and alumina creates a potential dust problem, which is readily controlled by engineering methods. The alkaline solution poses problems that may range from simple dermatoses to serious chemical burns. Rigorous housekeeping procedures must be established, personal protective equipment worn, and personal cleanliness encouraged.

The alumina is converted to aluminum in a reduction cell shown in Figure

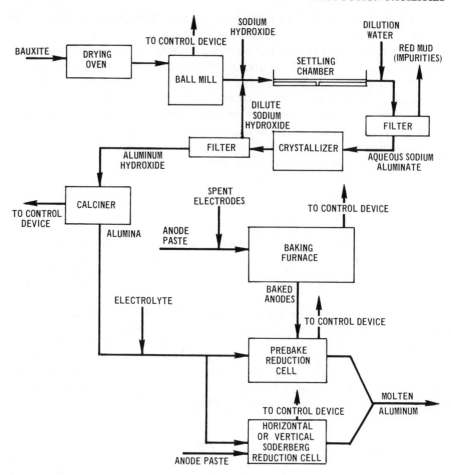

Figure 3.3-1 Schematic diagram of primary aluminum production process. (From Reference 2)

3.3-2 (2). A number of these cells or pots are arranged in lines in a potroom. The alumina is dissolved in cryolite (Na_3AlF_6); at the operating temperature of 980°C (1800°F) cryolite can dissolve up to 20% of alumina. Fluorspar (CaF_2) is added to the bath to lower the melting point of the mix and aluminum fluoride (AlF_2) to increase the efficiency of the cell. The steel cell has an inner lining of carbon that acts as the cathode and a consumable anode is made of petroleum coke and pitch. The low voltage (4–6 V)-high amperage (12,000–15,000 A) direct current passing through the cryolite-alumina bath reduces the alumina to aluminum metal and oxygen. The molten metal pooling at the bottom of the cell is siphoned off periodically, alumina is fed to the cell from hoppers, and cryolite and the consumable anodes are replaced. For each ton of aluminum metal produced, approximately 4 tons of dried bauxite are required, or 2 tons of Al_2O_3. The oxygen released at the anode forms carbon monoxide and carbon

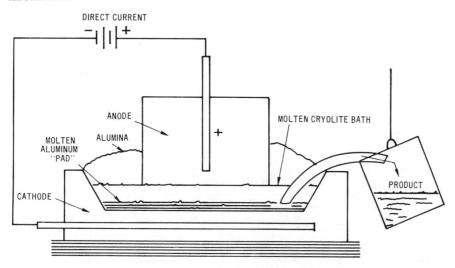

Figure 3.3-2   Aluminum reduction cell. (From Reference 3)

dioxide. In addition to CO and CO_2, the aluminum reduction cell releases polynuclear aromatic hydrocarbons from the consumable anode, fluoride compounds from the cryolite, fluorspar, and aluminum fluoride, and particulates from the handling of various granular materials.

The cell can be operated with either a prebaked anode or an anode material that cures in place. The Soderberg pot described above is shown in detail in Figure 3.3-3 and the prebaked anode cell in Figure 3.3-4. In the prebaked

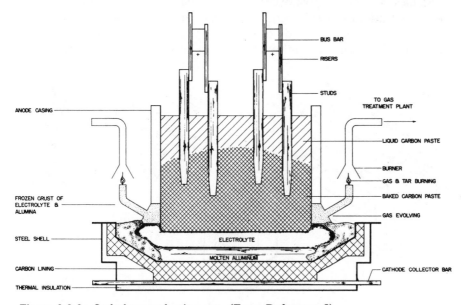

Figure 3.3-3   Soderberg reduction pot. (From Reference 5)

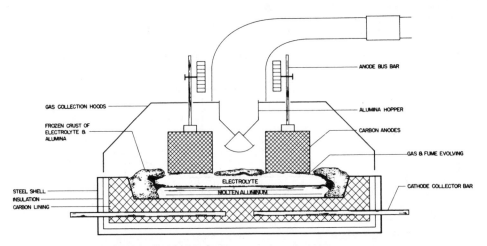

Figure 3.3-4 Prebake aluminum reduction cell. (From Reference 5)

process the anodes are fabricated in an auxiliary building at the aluminum reduction facility, as shown in Figure 3.3-5 (3). The anodes are made by grinding the petroleum coke and pitch, casting it in a form, and baking it until it has mechanical strength. During this operation the workers are exposed to dust with a significant concentration of benzene-soluble particulates. In the Soderberg anode pot system, the anode paste is baked in place by convective heat from the bath. In this procedure the volatiles are released to the potroom as the electrode "cures."

Studies of worker exposure to fluorides in aluminum production show that the concentration of gaseous and particulate fluorides are lowest in the prebake process (4, 5). A recent study of four aluminum reduction facilities in the United States has shown significant benzene-soluble particulate concentrations

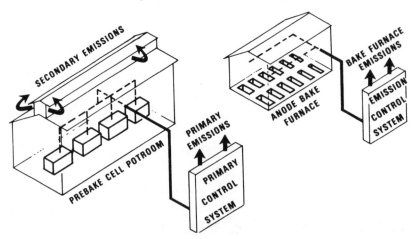

Figure 3.3-5 Prebake plant with anode bake furnace. (From Reference 3)

in potrooms, with the highest level found in the Soderberg process (5). The time-weighted average exposure for fluorides in these plants was below 2.5 mg/m³, and the carbon monoxide concentrations were not significant. Noise and heat stress are potential problems in potrooms and warrant study, as does the dermatitis problem from handling petroleum coke and pitch.

The principal control for air contaminants in potrooms is effective local exhaust ventilation on the pots as shown in Figures 3.3-6 and 3.3-7 (3). See pages 148, 149.

In general, the techniques used in subsequent aluminum fabrication operations do not cause unique occupational health problems. For example, the casting of aluminum alloys presents no problems more unusual than those that would be encountered in other nonferrous foundry operations. In welding on aluminum, however, the spectra generated by inert gas shielded arc welding procedures are unique and result in a rich UV spectrum and high ozone concentrations.

There are few data to support the claim that aluminum dust generated during machining operations causes either pneumoconiosis or systemic toxic effects. However, aluminum dust in the air may present an explosion hazard, and of course this material has been used in the manufacture of pyrotechnics.

REFERENCES

1   K. R. Van Horn, Ed., *Aluminum,* Vols. 1–3, American Society for Metals, Metals Park, OH, 1967.

2   "Compilation of Air Pollution Emission Factors," 2nd ed., U.S. Environmental Protection Agency, Publication No. AD-42, Research Triangle Park, NC, 1973.

3   EPA Report No. EPA-450/2-74-02a, "Background Information for Standards of Performance: Primary Aluminum Plants, Vol. 1, Proposed Standards," Environmental Protection Agency, Washington, D.C., 1974.

4   N. L. Kaltreider, M. J. Elder, L. V. Cralley, and M. O. Colwell, *J. Occup. Med.,* **14**, 531 (1972).

5   P. J. Shuler and P. J. Bierbaum, "Environmental Surveys of Aluminum Production Plants," U.S. Department of Health, Education and Welfare, Publication No. (NIOSH) 74-101, Cincinnati, OH, 1974.

3.4 AMMONIA

In the early 1970s the production of ammonia consistently ranked second or third on the list of top chemicals and first in the value of product. It is produced by 80 companies in over 100 U.S. plants. The major market for ammonia is fertilizer (80–85%), the second minor market is fibers and plastic intermediates (5%), and the balance is used in explosives, metallurgy, pulp and paper, and a variety of miscellaneous applications.

A number of processes are utilized to manufacture ammonia. All processes are based on the catalytic formation of ammonia from hydrogen and nitrogen at temperatures of 400–600°C (750–1110°F) and pressures of 10-90 MPa. A typical

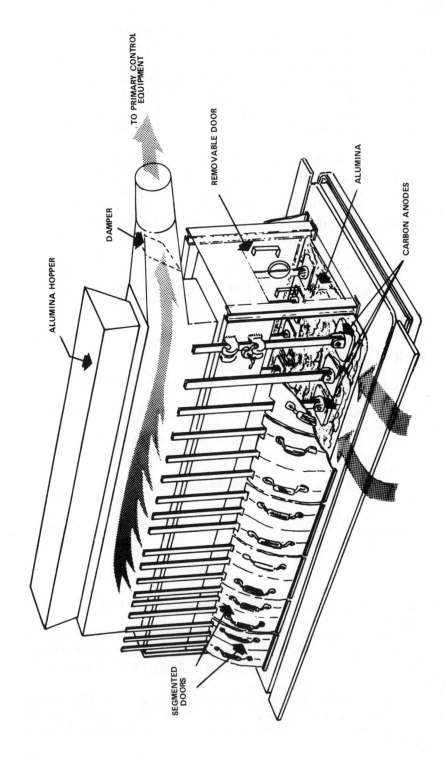

TO PRIMARY CONTROL EQUIPMENT

REMOVABLE DOOR

DAMPER

ALUMINA HOPPER

SEGMENTED DOORS

ALUMINA

CARBON ANODES

Figure 3.3-6 Typical prebake cell hooding. (From Reference 3)

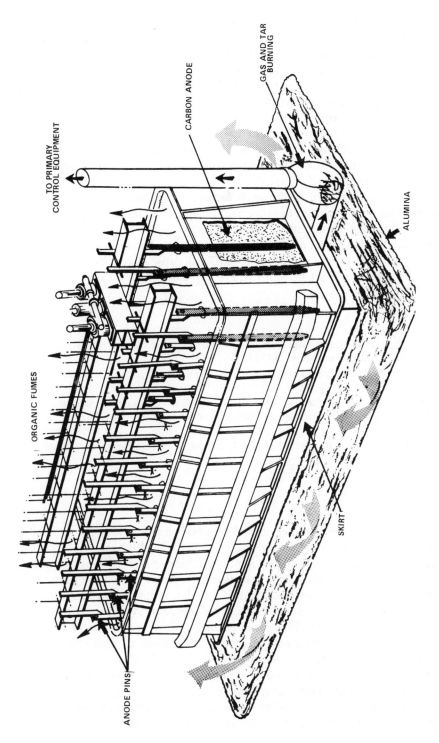

Figure 3.3-7 Typical vertical stud Soderberg cell hooding. (From Reference 3)

149

plant shown in Figure 3.4-1 utilizes natural gas as the source of hydrogen; nitrogen is obtained from the air. After desulfurization, the natural gas is refined over a nickel catalyst with steam to produce hydrogen and CO. In a secondary reformer, the CO reacts with steam to produce hydrogen and CO_2. The steam from the secondary reformer goes to a two stage shift converter that removes CO and produces hydrogen. The converter utilizes an iron catalyst in the first stage and a copper catalyst in the second stage. The CO_2 in the stream is removed by triethanolamine, organic solvents, or hot carbonates. This stream, containing 0.1% CO and 0.5% CO_2, is converted to methane before passing to the converter where hydrogen and nitrogen, in a 3-to-1 mole ratio, are converted to NH_3.

The principal hazard in the manufacture of ammonia is the accidental release of major amounts of the very irritating gas from failure of piping due to poor maintenance. Emergency escape and entry respiratory protection should be provided at ammonia plants. Fire and explosion hazards are a significant problem, and gas-to-air mixtures of 16–25% can cause violent explosions. Skin and eye protection must be worn where exposure may occur (1, 2, 3, 4).

REFERENCES

1   Chemical Manufacturer's Association, "Anhydrous Ammonia," Chemical Safety Data Sheet No. SD-8, CMA, Washington, D.C., 1960.
2   American National Standard Institute, "Standard for the Handling and Storage of Anhydrous Ammonia," K61.1-1972, ANSI, New York, 1972.
3   "Criteria for a Recommended Standard—Occupational Exposure to Ammonia," U.S. Department of Health, Education and Welfare, Publication No. (NIOSH) 74-136, Cincinnati, OH, 1974.
4   Fertilizer Institute, "Operational Safety Manual for Anhydrous Ammonia," Washington, D.C., 1978.

3.5 ARTWORK

The introduction of new materials and processes has provided the artist with new modes of expression, and unfortunately, new health hazards (1, 2). The painter has a number of synthetic paint media, based on acrylics, vinyl acrylics, vinyl acetate, and pyroxlyin. The vehicles and solvents now in use present the normal hazards of primary irritation, defatting dermatitis, and systemic toxicity due to inhalation. The advent of airbrush techniques and protective sprays for the completed art present high air concentrations of contaminants for short periods. Use of silk screen techniques can also result in significant solvent evaporation and exposure to the artist.

The control measures required for these materials include good housekeeping, personal cleanliness, and proper exhaust ventilation. Siedlecki recommends the use of disposable covers for work surfaces that may become contaminated, disposable gloves to avoid skin contact, and the use of mild soap

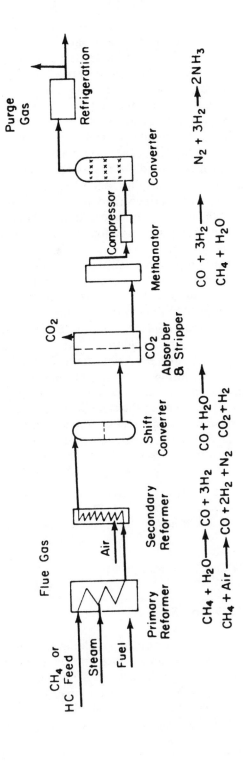

Figure 3.4-1 Typical plant for producing ammonia.

and water to remove resins and curing agents from the skin (1). Proper ventilation control may be difficult for the artist to achieve. In the studio of a firm employing several artists, it is feasible, but for the one-artist studio, respiratory protection may be a necessary substitute for ventilation.

The stone sculptor has also adopted new materials and processes used in industry. The dust exposure from working conventional materials such as marble and granite has been extended by high energy power tools, which can remove large quantities of material in a short time but generate very high dust concentrations in the process. In addition, new techniques such as flame cutting present serious hazards from fume, dust, and noise. The introduction of metal welded sculpture brought with it a new set of problems including UV radiation and exposure to metal fume, ozone, and nitrogen dioxide as described in Section 2.15. Structural resin systems now found in the sculptor's studio include urethane, polyester, acrylic, and epoxy resins. Some of these resin components can cause dermatitis and respiratory sensitization. All the important casting systems discussed in Section 2.6 are now used by the sculptor. The popularity of ceramics has introduced the hazards noted in Section 3.24.

In reviewing the range of health hazards encountered in the arts and crafts, McCann notes that a national conference on this topic included case histories on pottery, stained glass, silver soldering, furniture stripping, and stone sculpture (3). It is obvious that the health hazards faced by the artist now parallel those noted in the various industrial processes described in Chapter 2. An inventory of art materials and processes used in the artist's studio and teaching facilities must be made in the same manner as in industry (4). The recommendation for controls frequently is more difficult in the studio than in the industrial setting due to intermittent exposure and limited resources; however, the same principles apply.

REFERENCES

1 J. T. Siedlecki, *J. Am. Med. Assoc.*, **204**, 1176 (1968).
2 M. McCann, *Health Hazards Manual for Artists*, Foundation for the Community of Artists, New York, 1980.
3 M. McCann, *Art Hazards News*, **2**, 1 (1978).
4 "Health Hazard Evaluation Determination Report No. 75-12-321," U.S. Department of Health, Education and Welfare, NIOSH, Cincinnati, OH, 1976.

3.6 ASBESTOS PRODUCTS

The widespread application of asbestos to building materials, pipe insulation, fireproofing, textiles, and friction products makes it difficult to summarize completely the potential health hazards from this material. Hopefully, the attention given this substance by occupational health researchers will supplement this brief coverage. The types of asbestos used commercially are shown in Table 3.6-1. Chrysotile is the most widely used type, representing more than 90% of the total use.

Asbestos ore deposits contain from 3 to 30% asbestos. The crushing operations and the various milling operations are shown in Figure 3.6-1 with the major dust generation points that require local exhaust ventilation (1).

Table 3.6-1 *Types of Asbestos*

Chrysotile
Amphiboles
   Amosite
   Anthophyllite
   Crocidolite
   Tremolite
   Actinolite

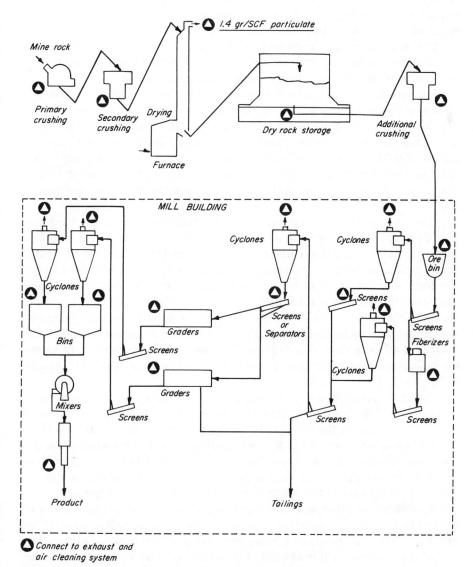

Figure 3.6-1 Flow diagram for milling of asbestos ores. (Courtesy of American Conference of Governmental Industrial Hygienists)

After primary and secondary crushing by jaw and cone crushers, the stock is dried and stored for processing. In the mill the rock is taken from storage bins and the fibers are released by mechanical attrition in fiberizers. The stock is then screened, and the airborne mix of fibers and granular material is conveyed to a cyclone where the fibers are separated from the stream and collected as "floats." The coarse material collected by the cyclone is classified by fiber length in graders.

In the initial phases of the mill operation, the bulk material is transferred by mechanical equipment including conveyor belts and skip hoists. Rigorous ventilation control is required at all material transfer points to establish control (2, 3, 4, 5). Once the fibers are released, the product flow is achieved by pneumatic conveying, and off streams, including cyclone exhausts, represent potential dust generation points.

The average air concentrations of fibers in a six mill study were 11.0 fibers/ml for crusher and dryer operators, 11.5 fibers/ml for mill operators, and 11.3 fibers/ml for blender, bagger, and packer workers (6).

The asbestos textile operations in Figure 3.6-2 are similar to those of the cotton industry and include carding, spinning, twisting, weaving, and braiding (7). Since the operations are conducted dry with relatively friable fibers, significant dust concentrations may be encountered. The conventional exhaust is typified by the complete enclosure of the card machine shown in Figure 3.6-3 (8). Dust concentrations noted in a modern United Kingdom textile facility ranged from 0.5 to 6 fibers/ml with carding and spinning the dustiest operations (9).

A second common use of asbestos is in the manufacture of asbestos-cement pipe and panels. These operations are characterized by the mixing of asbestos, silica, and cement as a slurry that is poured on mandrels or platens where the excess moisture is removed by vacuum. The wet product shape is removed from the form and air-cured at an elevated temperature. A conventional pipe manufacturing process that has all the elements of this type of manufacture is shown in Figure 3.6-4 (10). The only dry operations, the initial debagging and mixing and the final finishing or machining operations, represent the significant sources of asbestos dust. A NIOSH study of seven asbestos-pipe plants shows the highest mean concentration of 6.3 fibers/ml to be in the mixing operation (11).

One of the principal uses of asbestos is in the manufacture of vehicle-brake friction pads. This operation, shown in Figure 3.6-5, again presents an initial dust exposure during the mixing of asbestos and the other components (10). The pad stock is formed in presses and initially cured. It is cut to shape, and a final cure is achieved in an oven. The second major source of asbestos dust exposure occurs during the final finishing operations when the product is ground to dimension and various machining operations, including drilling and countersinking, are completed on the product. As one would anticipate, the highest asbestos fiber concentrations have been noted in cutting and drilling operations (14.4 fibers/ml) and during the mixing and preparation at the front end of the process (11.0 fibers/ml) (11). Again, exhaust ventilation must be

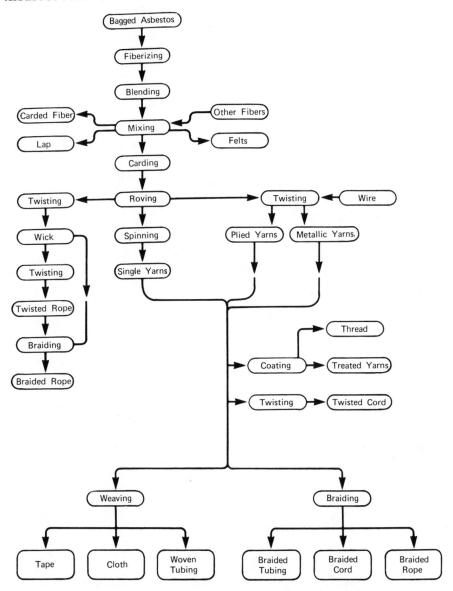

Figure 3.6-2 Flow diagram for asbestos textile manufacture. (From Reference 7)

applied to such operations as debagging, mixing, and machining. The control of asbestos dust can be accomplished by proper choice of manufacturing method coupled with effective local exhaust ventilation, air cleaning, and work practices. A compilation of guidelines for use in designing local exhaust ventilation systems has been prepared by Rajhans and Bragg (4).

The field use of these asbestos products is a matter of great concern. Installation of brake linings at a small garage facility usually is uncontrolled and

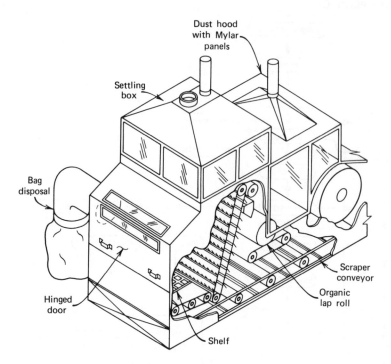

Figure 3.6-3 Textile card machine feed box. (From Reference 8)

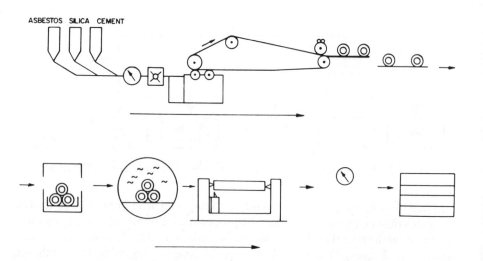

Figure 3.6-4 Flow diagram for asbestos pipe manufacture. (From Reference 10)

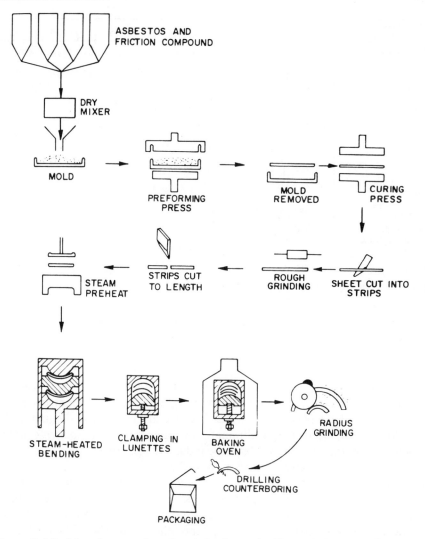

Figure 3.6-5 Manufacture of automatic brake pads. (From Reference 10)

can result in short-term exposure to high levels of asbestos. The field installation of asbestos insulation at stationary power plants and on shipboard also is difficult to control. In these cases, the control of asbestos taxes the ingenuity of the industrial hygienist. In many situations control can be achieved by substituting other materials including fibrous glass and ceramic insulation for asbestos. The suitability of asbestos substitutes depends on their temperature application range as noted in Figure 3.6-6 (12). Product design that minimizes the need for dry forming and machining operations is also of importance.

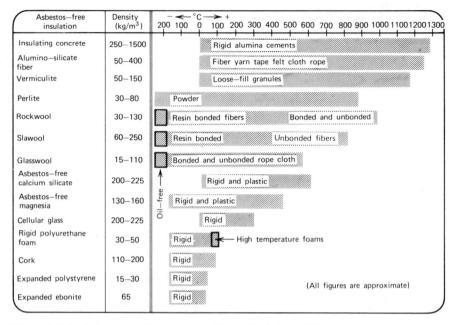

Asbestos—free insulation	Density (kg/m³)	− ← °C → +
Insulating concrete	250–1500	Rigid alumina cements
Alumino–silicate fiber	50–400	Fiber yarn tape felt cloth rope
Vermiculite	50–150	Loose–fill granules
Perlite	30–80	Powder
Rockwool	30–130	Resin bonded fibers / Bonded and unbonded
Slawool	60–250	Resin bonded / Unbonded fibers
Glasswool	15–110	Bonded and unbonded rope cloth
Asbestos–free calcium silicate	200–225	Rigid and plastic
Asbestos–free magnesia	130–160	Rigid and plastic
Cellular glass	200–225	Rigid
Rigid polyurethane foam	30–50	Rigid — High temperature foams
Cork	110–200	Rigid
Expanded polystyrene	15–30	Rigid (All figures are approximate)
Expanded ebonite	65	Rigid

Figure 3.6-6 Substitutes for asbestos. (From Reference 12)

The introduction of new techniques for doing a job also can be effective—for example, the use of a preweighed asbestos-cement insulation blend packaged in a plastic bag. Water is added at the job, and the material is hand kneaded. This technique eliminates the open mixing of asbestos cement, which is one of the very dusty jobs in the industry. The introduction of portable low volume-high velocity exhaust systems has made possible the control of asbestos dust from cutting asbestos pipe cover and pads in field operations. Housekeeping and rigorous personal hygiene, including the use of clothing change and showering, are also necessary to control asbestos. Vacuum cleaning should be used in place of dry sweeping, and other trades should be prohibited from the workspace during active insulation work.

A greater problem exists in the removal of installed asbestos material. Asbestos fibers are released to the air when vinyl-asbestos tile is sanded to provide a bonding surface before recovering. The application of sprayed asbestos for fireproofing structural steel in the 1960s has resulted in the covering of ceilings and walls with a loosely bonded asbestos blanket that can release fibers. If this material is applied to the air supply plenum surfaces in a building, it may result in a low level asbestos exposure to occupants. Removal of the material, encapsulation, or mechanical enclosure is difficult and expensive, but may be necessary. The removal or "rip-out" of degraded insulation from high pressure steam lines can also result in significant short-term exposures. Wet methods of removal are helpful in controlling this problem (13).

REFERENCES

1   Committee on Air Pollution, American Conference of Governmental Industrial Hygienists, *Process Flow Diagrams and Air Pollution Emission Estimates*, ACGIH, Cincinnati, OH, 1973.

2   "Control of Asbestos Dust," Technical Data Note 35, Department of Employment, HM Factory Inspectorate, London, December 1972.

3   "Recommended Code of Practice for the Handling and Disposal of Asbestos Waste Materials," Asbestos Research Council, London, 1973.

4   G. S. Rajhans and G. M. Bragg, *Engineering Aspects of Asbestos Dust Control,* Ann Arbor Science, Ann Arbor, MI, 1978.

5   Committee on Industrial Ventilation, American Conference of Governmental Industrial Hygienists, *Industrial Ventilation: A Manual of Recommended Practice*, 16th ed., ACGIH, Lansing, MI, 1980.

6   L. A. Schutz, W. Bank, and G. Weems, "Airborne Asbestos Fiber Concentrations in Asbestos Mines and Mills in the United States," Unites States Bureau of Mines, Technical Progress Report 72, July 1973.

7   *Handbook of Asbestos Textiles,* 3rd ed., Asbestos Textile Institute, Pampton Lakes, NJ, 1967.

8   J. Goldfield and F. Brandt, *Am. Ind. Hyg. Assoc. J.,* **36**, 799 (1974).

9   W. J. Smither and H. C. Lewinsohn, "Asbestosis in Textile Manufacturing," *Biological Effects of Asbestosis,* International Agency for Research on Cancer, Lyon, 1973.

10   G. Slwis-Gremer, *Ann. N.Y. Acad. Sci.,* **132**, 215 (1874).

11   "Criteria for a Recommended Standard—Occupational Exposure to Asbestos," U.S. Department of Health, Education and Welfare, Report HSM 72-10207, NIOSH, Cincinnati, OH, 1972.

12   *Recommendations for Handling Asbestos,* Engineering Equipment Users Association Handbook No. 33, EEUA, London, 1969.

13   J. Fontaine and D. Trayer, *Am. Ind. Hyg. Assoc. J.,* **36**, 126, (1975).

3.7   ASPHALT PRODUCTS

The term "asphalt" applies to a naturally-occurring deposit or a product recovered from crude petroleum by a refinery process. The material can be separated into three groups: asphaltenes, resins, and oils. The resins form a cover over the particulate asphaltenes, and the paraffinic, naphthenic, or naphtha-aromatic oils are the suspending oils. The great bulk of asphalt is now obtained from petroleum, and the refinery process permits production of material ranging from viscous liquids to solids.

In the United States, straight petroleum asphalt is used; in Europe, coal tar pitch is used to "cut" the petroleum-based asphalt. The difference in toxicity between these asphalts is significant (1). The petroleum asphalt is not considered highly toxic. The principal health problems appear to be thermal burns from skin contact and possibly skin sensitization. The carcinogenicity of petroleum asphalt has not been demonstrated in human beings. European asphalt containing coal tar pitch, however, presents a potential hazard to polycyclic aromatic hydrocarbons and warrants entirely different controls due to its possible carcinogenicity.

A summary of handling procedures for asphalts based on petroleum asphalt "cut" with coal tar pitch from Sanderson and Jelfs (1) is quoted later. Although the level of hazard connected with asphalts in the United States is not as great as the hazard for European products, for which these guidelines were prepared, precautions seem appropriate for all asphalts based on present knowledge.

1 Avoid prolonged and repeated skin contact with bitumen materials.

2 Wear suitable protective clothing, especially gloves, and do not wear heavily soiled garments. Dry cleaning is recommended for soiled clothing.

3 Never put dirty rags or towels in pockets.

4 Remove bitumen contamination from the skin by thorough washing with skin cleanser and warm water. Wash hands thoroughly before and after going to the lavatory. Kerosene and other solvents must not be used for normal skin-cleansing purposes.

5 Obtain medical advice if any skin changes are noticed, particularly if these occur in areas that have been exposed to contact with bitumen/coal tar products.

3.7.1 Paving Plants

A major use of petroleum asphalt is in paving products for roads and runways. Since the final product cannot be shipped long distances, a large number of paving plants of the type shown in Figure 3.7-1 are scattered throughout the United States (2). Sand and gravel, the major aggregates, are quarried near the plant and dumped directly onto hoppers by truck or the aggregate is stockpiled and transferred to hoppers by crane or front-end loaders.

The granular material is dried in an oil-or gas-fired rotary drier. The hot aggregate is screened and held in a bin directly over the mill. When a truckload of asphalt is to be prepared, a given blend by weight of aggregate, filler such as fly ash, limestone or Portland cement, and hot asphalt from a heated asphalt storage tank is dumped into the mixer and blended for 1–2 min. The blended mix is dropped onto a truck for transport to the paving site.

During this sequence the operator of the plant may have a serious heat, noise, dust, and asphalt fumes exposure depending on the design of the facility and the availability of enclosed, conditioned control rooms. Equipment operators may be exposed to significant dust concentrations if they operate with open cabs. Maintenance personnel may have serious exposures to heat,

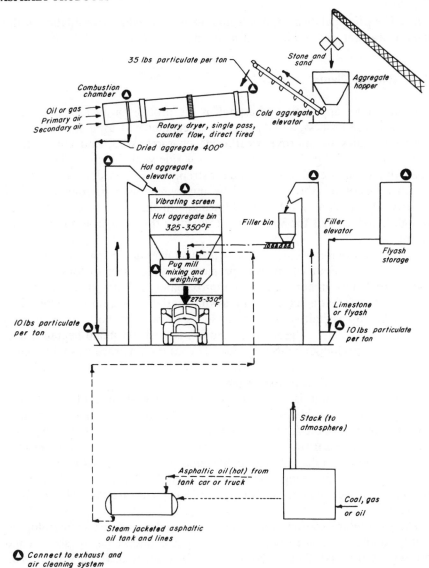

35 lbs particulate per ton

Stone and sand

Aggregate hopper

Combustion chamber

Oil or gas
Primary air
Secondary air

Cold aggregate elevator

Rotary dryer, single pass, counter flow, direct fired

Dried aggregate 400°

Hot aggregate elevator

Vibrating screen

Hot aggregate bin 325-350°F

Filler bin

Filler elevator

Flyash storage

Pug mill mixing and weighing

275-350°F

Limestone or flyash

10 lbs particulate per ton

10 lbs particulate per ton

Stack (to atmosphere)

Asphaltic oil (hot) from tank car or truck

Coal, gas or oil

Steam jacketed asphaltic oil tank and lines

⬤ Connect to exhaust and air cleaning system

Figure 3.7-1   Flow diagram for asphalt paving plant. (From Reference 2, courtesy of American Conference of Governmental Industrial Hygienists)

noise, dust, and asphalt fume exposure depending on the design of the facility equipment during plant operation. The health hazard from the exposure to mineral dust depends on the free silica content of the rock and sand. Air concentration data are not available in the open literature on asphalt paving plants. The principal dust control technique is local exhaust ventilation at the material transfer points identified in Figure 3.7-1 (3).

3.7.2 Roofing Plants

A second major application of asphalt is the production of asphalt roofing products. In this process, felted paper is impregnated with asphalt, coated with mineral granules to improve weathering characteristics, and cut to form conventional roof shingles, roll covering, or side wall shingles. The felt is a paper with wood or cloth fiber added for strength; fire retardant shingles may contain asbestos or fibrous glass. The impregnant is a petroleum asphalt prepared at the plant site by "air blowing." In this process air is bubbled through liquid asphalt for 8 to 16 hrs to oxidize the asphalt and increase its melting point.

A schematic of a continuous roofing product manufacturing line is shown in Figure 3.7-2 (4). Four component preparation lines meet to form the final product: rock dust, mineral granules approximately 1–2 mm in diameter, felt, and asphalt. The felt is fed from an unwind station to a saturator where it is impregnated by asphalt spray, dip, or in some cases by both techniques as shown in Figure 3.7-2. The impregnated felt is dried on steam heated drums and is fed to a wet looper or accumulator. The saturation process usually results in worker exposure to asphalt fumes. Control normally is achieved by an enclosed canopy with roof exhauster.

The warm asphalt is then "filled" with mineral dust obtained as a by-product during crushing of the granules. Talc, soapstone, mica, or sand is applied to the rear of the felt to keep it from sticking. The filled felt then passes to a granule applicator station where pigmented granules are pressed into the dust-filled asphalt on the felt. These two operations may be dusty and involve dust exposure to the range of minerals used in the coater and granule applicator. After cooling, the continuous roll is cut to form either roll roofing or shingles.

The acceptable exposure limit to mineral dust depends on its mineralogical identity and its crystalline silica content. If special fillers such as asbestos and fibrous glass are used, these exposures must be evaluated. If talc is used as an antisticking agent on the rear of the product, it should be evaluated to ensure that it is nonfibrous and to determine its free silica content. Nonasbestos talcs with less than 1% free silica are available for this application. The adhesive used to seal one shingle to another on the roof is applied in the plant and its composition should be identified. In many cases it is merely a dab of asphalt.

In a NIOSH Health Hazard Evaluation of an asphalt shingle plant, air samples were collected for asbestos, crystalline silica, formaldehyde, asphalt, aromatic hydrocarbons, and total aliphatic hydrocarbons (5). The percentage of crystalline quartz in respirable dust samples taken on various operations in

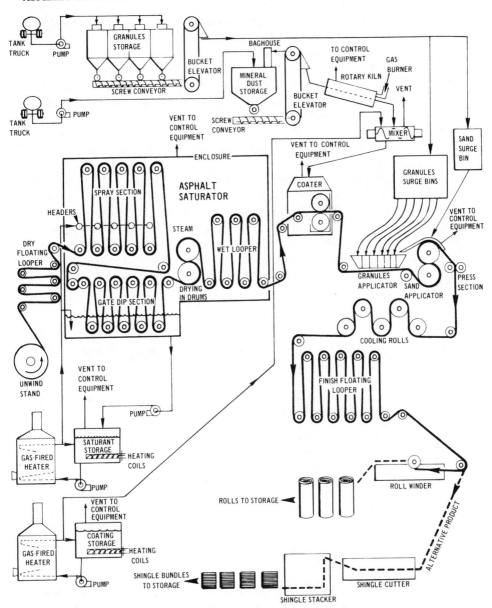

Figure 3.7-2 Schematic of a continuous roofing product manufacturing line. (From Reference 4)

this roofing plant ranged from 0 to 33.3%; the percentage in the total dust samples was 49.8 to 55.5%. The respirable dust concentrations on the granule application and tab seal operations varied from 0.13 to 0.20 mg/m³ and, for total dust, 0.89 to 3.1 mg/m³. Asphalt fume concentrations on these operations were 3.25 and 2.10 mg/m³; however, the authors stated that additional sampling demonstrated serious inconsistencies in the sampling procedure. Aromatic hydrocarbons were not detected and total aliphatic hydrocarbon (C7 to C12) concentrations on the coaters were 3.4 and 5.2 mg/m³ and on the saturator coating, 4.8 and 25.4 mg/m³. The authors state that polynuclear aromatic compounds in petroleum-based asphalts are highly variable and found only in trace amounts. Benz(a)pyrene (BAP) and benz(a)anthracene were found in microgram per cubic meter levels in the saturator. BAP was found in two out of seven samples in the general work area (0.35 μg/m³ and 0.10 μg/m³).

Asphalt processes in roofing plants release dense white clouds of hydrocarbon oils, which must be controlled by local exhaust ventilation at the asphalt storage tanks, saturator, drying drums, and wet looper. Dust from the sand, mica, and talc used in the process may require ventilation control at the coater and granule applicator. The saturators, drying-in drums, and wet loopers are usually ventilated by enclosure-type hoods Air cleaning in the form of incinerators and scrubbers is required.

REFERENCES

1 J. T. Sanderson and E. C. G. Jelfs, *Med. Bull. Exxon Corp. Affil. Co.*, **38**, 27 (1978).

2 Committee on Air Pollution, American Conference of Governmental Industrial Hygienists, *Process Flow Diagrams and Air Pollution Emission Rates*, ACGIH, Cincinnati, OH, 1973.

3 Committee on Industrial Ventilation, American Conference of Governmental Industrial Hygienists, *Industrial Ventilation: A Manual of Recommended Practice*, 16th ed., ACGIH, Lansing, MI, 1980.

4 J. Danielson, Ed., *Air Pollution Engineering Manual*, 2nd ed., Publication No. AP-40, Government Printing Office, Washington, D.C., 1973.

5 Health Hazard Evaluation Determination Report No. 77-56-467, Department of Health, Education and Welfare, NIOSH, Cincinnati, OH, 1978.

3.8 BATTERIES

3.8.1 Lead-Acid Batteries

The charged battery cell consists of a series of plates: the negative electrode is in the form of spongy metallic lead and the positive plate is lead dioxide (PbO_2). The electrolyte for this battery system is an aqueous solution of sulfuric acid. The equation of the reaction is

$$2Pb + 2PbO_2 + 2H_2SO_4 + H_2O \rightarrow 2PbSO_4 + 2PbO + 3H_2O$$

During charging the negative plate changes from lead sulfate ($PbSO_4$) to spongy lead, while the positive electrode is oxidized from lead monoxide (PbO) to lead dioxide (PbO_2). The battery discharges with a characteristic voltage of 2.0 V, as shown in the preceding equation.

The battery plant is normally divided into lead oxide production, pasting, finishing, and forming areas. Each production area has a characteristic exposure pattern; however, in small plants the manufacturing sections adjoin one another and the workers normally rotate through all jobs. The main production activities are shown in Figure 3.8-1 and are discussed later (1).

Grid casting

Molten lead alloy at 370°C (700°F) is fed to the grid casting machine where it flows to the individual mold cavities, cools, and solidifies to form the grid. To improve characteristics, lead-antimony alloys ranging from 3.5–7.0% antimony are used as grids, connectors, and posts in lead-acid batteries. The excess lead or flashing is trimmed from the part and returned to the lead melting pot. Since the vapor pressure of lead is low at its melting point, it is not the major

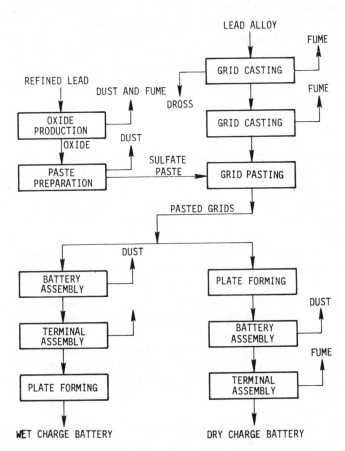

Figure 3.8-1 Flow diagram of a lead-acid battery plant. (From Reference 1)

source of lead fume; however, the oxidation products on the surface of the pot must be removed and this drossing operation may result in serious exposure.

Paste production and application

In small plants, the lead monoxide used in the manufacture of battery plates is purchased from an outside supplier. In large plants, it is prepared on site usually by one of two methods. In the first process, molten lead is oxidized by an air blast and subsequently ground and sized before being stored for plant use. In the second method, the pig lead is held in a reverberating furnace at controlled temperature to form the oxide. These facilities must be provided with local exhaust ventilation and suitable air cleaning. The lead monoxide is mixed with water, sulfuric acid, and other additives depending on whether positive or negative plates are being produced. This work must be conducted with local exhaust ventilation at the location where the lead oxide is charged to the mixer.

The cast grids are hand-fed to a belt conveyor that passes the plates under a hopper containing the moist paste. The voids in the grids are filled with paste, the paste is pressed in place, and the plates are passed through a low temperature drying oven. Until the grids or plates enter the oven, the paste is moist. After drying, the material is friable and represents the principal source of lead dust in the workroom. The pasting machine normally is not provided with local exhaust ventilation.

Finishing and forming

When the plates leave the oven, they are cleaned with a power wire brush to remove excess plate debris and clean the tabs. This is a dusty operation and warrants close attention.

In the production of dry-charge batteries the plates are stacked, electrically connected, and "formed" electrically in a dilute H_2SO_4 bath so that the $PbSO_4$ (anode) is converted to Pb and the negative electrode PbO plate is converted to PbO_2. Then the assembly is placed in a battery case, which is usually cast from hard rubber or plastics such as polystyrene, polypropylene, polyvinyl chloride, or acrylonitrile-butadiene-styrene. Coal or petroleum pitch, asbestos, or other minerals may serve as fillers. Finally the top of the battery is sealed with asphalt or epoxy resin. In the wet-charge battery, the battery is assembled, electrolyte added, and the plates are "formed."

The lead smelting and reclamation operations associated with battery manufacture present a critical problem and require rigid ventilation control and good housekeeping. The oxide mixing facility must be equipped with a dust control system, and dispensing operations should be designed to minimize exposure to lead.

From the time the lead oxide paste is applied to the plates, the control of the exposure becomes more difficult. The application itself presents little difficulty, but as soon as the paste becomes dry it is easily dispersed into the air.

This may occur when the plates are subsequently cleaned, transferred, racked, stacked, trimmed, and split. If the wet paste is allowed to fall on the floor or contaminate equipment, sooner or later it will dry and add to the airborne dust. Good housekeeping is absolutely essential in this industry.

Work tables with downdraft exhaust have proved effective in controlling dust release, as well as floor contamination and resuspension. The plates should be handled manually as little as possible. The plates of ready-charged batteries are handled and shipped dry, ready for service as soon as the electrolyte is added, and these offer more serious dust-control problems than plates that are handled and shipped wet.

Controlling lead exposure in storage battery manufacture is not a piecemeal problem, but one that requires a well-rounded health maintenance program, including (a) competent engineering control, (b) preemployment examinations with biological monitoring, (c) periodic surveys of atmospheric contamination, (d) periodic blood-lead determinations of persons in potentially hazardous workplaces, (e) education of employees about personal hygiene, good health practices, and safe working procedures, and (f) transfer of overexposed workers to a position of minimum exposure (2, 3).

Floors must be surfaced and covered for easy cleaning. Vacuum cleaning must be used; protective clothing and suitable eating and sanitary facilities divorced from the production area are necessary.

The battery charging room facility is subject to contamination by hydrogen and acid mist, and since battery grids often contain from 5 to 10% antimony, production of stibine is at least theoretically possible. Dilution ventilation is the conventional control technique in the charging area.

The air concentrations of lead and the corresponding blood concentrations on various jobs in a United Kingdom battery plant are shown in Table 3.8-1 (4). A 1977 survey of a large battery plant by NIOSH showed that grid mold, pasting, battery assembly, and lead reclamation areas had airborne levels that exceeded 100 $\mu g/m^3$. One quarter of the samples taken in grid molding, battery assembly, and reclamation exceeded 200 $\mu g/m^3$ (5).

Table 3.8-1 *Representative Mean Lead Exposures and Biologic Lead Levels for Workers in the Storage Battery Industry*

Job	Number of Workers	Mean Air Lead Concentration (mg/m^3)	Mean Blood Lead Concentration (g/100 g blood)
Machine pasting	6	0.218	74.2
Hand pasting	8	0.150	63.2
Forming	9	0.134	63.0
Casting	6	0.052	—
Plastics Dept. A	5	0.012	27.2
Plastics Dept. B	5	0.009	29.1

Source. Reference 4.

3.8.2 Nickel-Cadmium Batteries

In the nickel-cadmium battery, both negative and positive electrode materials are contained in porous plates. The negative electrode is cadmium hydroxide $(Cd(OH)_2)$ and the positive electrode is nickel hydroxide $(Ni(OH)_2)$; the electrolyte in this system is an aqueous solution of potassium hydroxide (KOH). The charge-discharge equation is

$$\begin{array}{c} \text{Discharge} \\ \rightarrow \\ Cd + 2H_2O + 2NiOOH \qquad\qquad 2Ni(OH)_2 + Cd(OH)_2 \\ \leftarrow \\ \text{Charge} \end{array}$$

When charged, $Cd(OH)_2$ is reduced to cadmium metal and $Ni(OH)_2$ is converted to NiOOH.

Cadmium and nickel plate materials can be manufactured in a number of ways and the hazards in their manufacture have not been published in the open literature. It is common to have the prepared active electrodes shipped to a plant for assembly of vented and sealed battery cells. Such a plant was surveyed by NIOSH in a Health Hazard Evaluation (6). In this plant the major exposures to nickel and cadmium are from the handling of nickel containers and the active ingredients on the plates. Cobalt hydroxide $(Co(OH)_2)$ may also be present in the positive electrode material. In this assembly plant the electrode stock is cut to size, terminals are welded to the plates, and the plates are interleaved with separators. The assembly is placed in a container, electrolyte is added, and the cells are sealed. Air concentrations of cadmium and nickel routinely exceeded 0.10 mg/m³ and 1.0 mg/m³, respectively.

The investigators recommended that cracker machines and terminal welding stations in this plant be provided with local exhaust ventilation. As in the case with lead-acid batteries, compressed air and sweeping should not be used for cleaning; rather, a central vacuum system should be utilized.

3.8.3 Dry Cell Batteries (Leclanche)

The negative electrode of the dry cell is the zinc can, which also serves as the structural container for the battery. The positive electrode is manganese dioxide (MnO_2) with an electrolyte consisting of an aqueous solution of ammonium chloride (NH_4Cl) and zinc chloride $(ZnCl_2)$, which infuses the manganese dioxide. An organic paste inner lining on the negative electrode also is saturated with NH_4Cl and $ZnCl_2$. This sleeve contains small quantities of mercuric chloride $(HgCl_2)$, which forms an amalgam on the inner surface of the zinc can. A carbon rod embedded in the positive electrode material acts as the current collector.

The discharge reaction of the battery is

$$2MnO_2 + 2NH_4Cl + Zn \rightarrow ZnCl_2 \cdot 2NH_3 + H_2O + Mn_2O_3$$

Heavy-duty zinc carbon cells have the same construction and reaction as the standard Leclanche cell except that the electrolyte is based solely on zinc chloride.

In the alkaline manganese cell, the negative electrode is finely divided zinc amalgamated with mercury. A highly conductive alkaline KOH electrolyte is used to impregnate the electrode mass. The positive electrode is based on an electrolytically-produced MnO_2. Electrolyte is also absorbed into the positive electrode mass and a metal current collector is immersed in the electrolyte. The charge-discharge reactions for this battery are shown as both a primary and rechargeable system.

Primary

$$2Zn + 2KOH + 3MnO_2 \rightarrow 2ZnO + 2KOH + Mn_3O_4$$

Rechargeable

$$Zn + KOH + 2MnO_2 \rightleftharpoons ZnO + Mn_2O_3 + KOH$$

The principal hazard in dry cell manufacture is attributed to MnO_2 dust generated from the initial mixing and later handling of the depolarizer mix. However, there can also be exposure to ammonium and zinc chloride during weighing, mixing, and dispensing. Once the electrolyte is added, the dust exposure is minimized. The use of mercury and its salts also presents a potential hazard and warrants air sampling and biological monitoring.

The controls required during preparation of the depolarizer include ventilation, wearing of personal protective clothing, and rigorous housekeeping based on vacuum cleaning of the area.

3.8.4 Mercury Cell

The mercury battery has a negative electrode of powdered zinc and a positive electrode of mercuric oxide (HgO). An aqueous solution of potassium hydroxide (KOH) is infused in both electrodes and the necessary separators. The discharge reaction is

$$Zn + HgO + KOH \rightarrow ZnO + Hg + KOH$$

The manufacturing process takes place in four specific phases. In the oxide plant HgO is formed by first cleaning the mercury and then combining it with chlorine and sodium hydroxide in enclosed reactors. The oxide is transferred from the reactors, dried, and packaged. The major exposures during this operation are to HgO dust and metallic mercury vapor.

The negative zinc-mercury amalgam is formed in the anode room and pelletized in the geometry required for battery assembly. The principal exposure occurs during handling and charging of the zinc-mercury materials with limited exposure during press operations.

Material for the positive electrode is made by mixing manganese oxide, graphite, and cadmium and pressing the mixture into the battery case. Exposure to mercury vapor and dust from the various electrode components occurs in the depolarizer room.

The two electrodes and associated piece parts are assembled mechanically. Exposure to mercury vapor is negligible in this area.

The principal exposures in this operation are to particulate mercury and elemental mercury. The range of airborne levels shown in Table 3.8-2 were noted by NIOSH in a Health Hazard Evaluation survey (7). These data indicate major exposures to mercury in the depolarizer room, anode room, and oxide plant. The air sampling data were supported by elevated mercury in urine data.

Table 3.8-2 *Air Concentrations of Mercury Vapor*

Location	Job	Average Exposure (mg/m³)	95% LCL–UCL (mg/m³)
Depolarizer room	Press operator	0.56	0.41–0.71
	Slugger operator	0.77	0.51–1.03
	Mechanic	0.50	0.26–0.72
Anode room	Amalgam blender	0.26	0.11–0.41
	Press operator	0.09	0.05–0.13
Oxide plant	Process operator	0.45	0.04–0.96
	Material handler	1.29	0.5–3.09

Source. Reference 7.

REFERENCES

1 "Control Techniques For Lead Air Emissions," Vol. 2, Chap. 4, Appendix B, U.S. Environmental Protection Agency, EPA-450/2-77-012, Research Triangle Park, NC, 1977.

2 S. Tola, S. Heinberg, J. Nikkanen, and S. Valkonen, *Work Environ. Health*, **8**, 81 (1971).

3 "Criteria for a Recommended Standard—Occupational Exposure to Inorganic Lead," U.S. Department of Health, Education and Welfare, Publication No. HSM 7-11010, Cincinnati, OH, 1973.

4 M. K. Williams, E. King, and J. Walford, *Br. J. Ind. Med.* **26**, 202 (1969).

5 "Health Hazard Evaluation Determination Report No. HE 77-28-552," U.S. Department of Health, Education and Welfare, NIOSH, Cincinnati, OH, 1977.

6 "Health Hazard Evaluation Determination Report No. 74-16-272," U.S. Department of Health, Education and Welfare, NIOSH, Cincinnati, OH, 1976.

7 "Health Hazard Evaluation Determination Report No. 78-26-560," U.S. Department of Health, Education and Welfare, NIOSH, Cincinnati, OH, 1979.

3.9 BERYLLIUM

Beryllium is used extensively in industry because of its unique properties of high tensile strength, workability, and electrical and thermal conductivity. The major industrial uses of beryllium and the principal exposures are given in Table 3.9-1. The exposures occur in the processing plant and among the intermediate producers and product manufacturers. It has been stated that there are 8000 plants handling beryllium in the United States and 30,000 workers with potential exposure. Controls on beryllium producers were established by the U.S. Atomic Energy Commission in 1949. Although the toxicity of this metal has been well described, new cases of disease continue to occur.

Table 3.9-1 *Air Contaminants Released During Beryllium Processing*

Process	Contaminant
Extraction plants	
Ore crushing, milling, mulling	Beryl ore dust
Briquetting, crushing, and milling	Briquette dust
Sintering	Beryl dust, sinter dust
Beryllium hydroxide production	$Be(OH)_2$ slurry, H_2SO_4 fume
Beryllium metal production	$(NH_4)_2BeF_4$, $PbCrO_4$, CaF_2, HF, $Be(OH)_2$, BeF_2, NH_4F, Mg, Be, MgF_2, BeO, acid fume
Beryllium oxide production	BeO fume and dust
Beryllium-copper alloy production	Be, Cu dust, BeO dust
Beryllium alloy machine shops	
Beryllium machining operations	Be dust
Beryllium-copper foundries	
Melting ingots in primary crucibles	Be fume
Preheating transfer crucibles	Be fume
Drossing and dross handling	Be fume
Pouring molds	Be fume
Finishing operations	Be dust
Beryllium ceramics	
Spray drying of oxide	BeO dust
Kilns	BeO dust, binders
Machining	BeO dust, binders

Source. Reference 1.

The major sources of beryllium emissions have been identified as beryllium metal extraction plants, ceramic production plants, foundries, machining facilities, and propellant manufacturing plants (1). The extraction plants produce beryllium metal, beryllium oxide powders, and copper, nickel, and aluminum alloys of beryllium. The varied exposure to dust, fumes, and mists of beryllium in extraction plants must be controlled by well-maintained local exhaust ventilation systems and efficient air-cleaning devices (2). The Beryllium Case Registry has recorded cases of berylliosis in smelting and extraction attributed to initial exposure after 1949, when control measures were adopted.

The production of beryllium ceramic has the greatest potential for exposure to workers from the varied operations shown in Figure 3.9-1 including milling, screening, drying, sintering, and machining. These operations must normally be conducted in glove boxes or total enclosures to ensure adequate control and, as shown in Figure 3.9-1, rigorous air cleaning is required.

Foundries take master alloys and recast them into special purpose beryllium-copper alloys of varying beryllium content. As in any other foundry job, exposure in these operations occurs as a result of fumes evolved during the melting and pouring of the alloy, and dust from drossing, shakeout, and metal finishing operations. In general, the conventional ventilation and air-cleaning

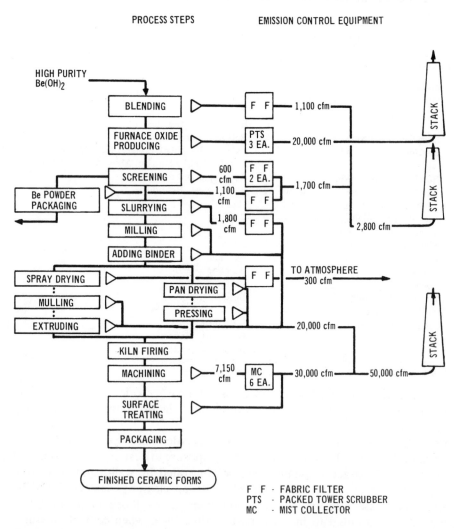

Figure 3.9-1 Conversion of beryllium hydroxide to beryllium oxide powder and ceramics. (From Reference 1)

techniques installed in foundries for metal fumes are not adequate to control the exposure to beryllium.

Breslin has suggested a format that is useful in determining the controls required on beryllium operations (3):

1 The beryllium operations should be segregated from all other plant work. Access should be limited, and the area should be adequately posted.

2 Process techniques that will minimize dust or fume release should be used, and the manual transfer of material discouraged.

3 Almost all processes must be ventilated, even though the process is a wet one. In the case of beryllium alloys, this may not be necessary; however, air samples should be taken on all operations to ensure that no hazard is presented. The hood and system design generally must be superior to that normally seen in industry. When possible, enclosing hoods should be used because they give superior performance. Small bench operations can be controlled with table top enclosure or glove boxes. Conventional machining operations have been controlled by enclosures or by low volume-high velocity exhaust systems.

4 The importance of work practices in control of beryllium exposure can be noted by monitoring different employees engaged in the same task. The care that one person may take in following specific task instructions can make a significant difference in his or her exposure.

5 Spills must be cleaned up immediately by vacuum systems. A high-efficiency filter must be used on the vacuum system discharge. Housekeeping must be encouraged by proper design of the facility, including continuous floor covering, coved corners and joints, and regular surfaces.

6 Personal protection normally includes frequent issue of personal protective clothing, which is to be worn only in the plant. Locker rooms should have "clean" and "dirty" sections to isolate effectively the production area and reduce clothing contamination.

7 Depending on the concentration, half- or full-facepiece air-purifying respirators, air line respirators, or self-contained breathing apparatus are required.

8 All controls must be monitored adequately by air sampling, and medical management of exposed workers is essential.

REFERENCES

1 "Control Techniques for Beryllium Air Pollutants," U.S. Environmental Protection Agency Publication No. AP-116, EPA, Research Triangle, NC, February 1973.

2 "Criteria for a Recommended Standard—Occupational Exposure to Beryllium," U.S. Department of Health, Education and Welfare, Report Publication No. (NIOSH) HSM 72-10268, Cincinnati, OH, 1972.

3 A. Breslin, in *Beryllium—Its Industrial Hygiene Aspects*, H. E. Stokinger, Ed., Academic, London and New York, 1966.

3.10 BRICK AND TILE

Various ceramic industrial products are manufactured in small facilities throughout the United States. The most common products are brick and tile; however, plants producing clay pipe and other construction materials have similar workplace exposures.

A flow diagram showing the common steps in the manufacturing process for bricks is presented in Figure 3.10-1 (1, 2). The principal raw materials are clay and shale, which are usually quarried in an open pit. The exposures in quarrying operations are described in Section 3.31 and include mineral dust and both hot and cold environmental extremes.

The bulk clay and shale are usually crushed initially at the mining site; further size reduction and classification are conducted at the plant. The secondary crushing and screening operations involve multiple transfer points that represent dust sources. In addition, unless maintenance is excellent, fugitive leaks from equipment will also contribute to worker dust exposures. The obvious control is local exhaust ventilation at all transfer points and a continuous preventive maintenance program. The importance of the dust exposure depends principally on the quartz content of the clay or shale.

The raw material is formed into brick by either a dry or wet process. In the dry process the fine granular material is slightly moistened and injected into molds operated at $3.5–12.4 \times 10^6$ Pa (500–1800 psig) (1). The resulting cohesive form is fired directly in a kiln. In the more common stiff-mud process, water is added to the clay in a pug mill to make a thick plastic mass. The mud is then extruded through a rectangular die. A wire cutter chops the continuous extrusion into brick length forms that are air dried and fired in a kiln. The forming and cutting operation involves a dust exposure during charging of the mill, but once the water is added dusting is minimal.

Commercial ceramic products are frequently glazed before firing. Glazing

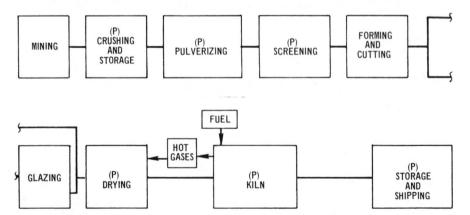

Figure 3.10-1 Flow diagram of brick manufacturing process. P denotes a major source of particulate emissions. (From Reference 1)

involves applying a slip to the surface that forms a smooth reflective skin after firing. The glazes may include heavy metal compounds.

The kilns are fired by gas or oil depending on the geographical region in which the plant is located. Firing of brick at maximum temperatures of 1090°C (2000°F) for 50 to 100 hr evaporates free water, dehydrates the mass, oxidizes available material, and vitrifies the product. At the kiln location, the major hazards are noise from the burners, heat stress principally from radiation, and exposure to fluorides from the raw materials and to carbon monoxide and sulfur dioxide from the products of combustion. No data are available on the exposure of operators in brick and tile plants.

REFERENCES

1 "Compilation of Air Pollution Emission Factors," 2nd ed., Publication No. AP-42, U.S. Environmental Protection Agency Research Triangle, NC, 1973.
2 "Technical Notes on Brick and Tile Construction," Structural Clay Products Institute Pamphlet No. 9, Washington, D. C., September 1961.

3.11 CEMENT

Cement is one of the essential materials of construction with plants sited throughout the United States. The raw materials consist of approximately 75% limestone, 20% clay, and 5% gypsum and iron. The final product has the following approximate composition: 75% calcium silicates, 5–10% calcium aluminates, 5% calcium sulfate, 1% oxides of sodium and potassium, 2–4% magnesium oxide, and 5–10% calcium-aluminum-iron compounds. As shown in Figure 3.11-1, after being quarried, the various minerals are crushed, proportioned by weight for a given product, and then calcined in a rotary kiln at approximately 1400°C (2550°F) (1). The clinker is cooled and mixed with gypsum to form the cement product, which is packaged for distribution.

The principal hazard in the operation of cement plants is dust and varies with the quartz content of the clay, shale, or slate used to form the clinker. Although the rock prepared for calcining may contain significant quantities of quartz, it is usually less than 1% in the finished product. Special products such as acid-resistant cements may use high concentrations of siliceous rock; some are also blended with asbestos to form refractory and insulation cements. Occasionally cement may contain diatomaceous earth, which, when calcined, forms cristobalite, a biologically active mineral.

The first step in evaluating the exposure of operators of a cement plant is to determine the mineralogical composition of the raw materials. If silica is present, air sampling must be carried out to demonstrate the respirable concentration of silica. The dust generating equipment, such as the crusher, grinder, and sizing screens, and the various granular transfer points must be

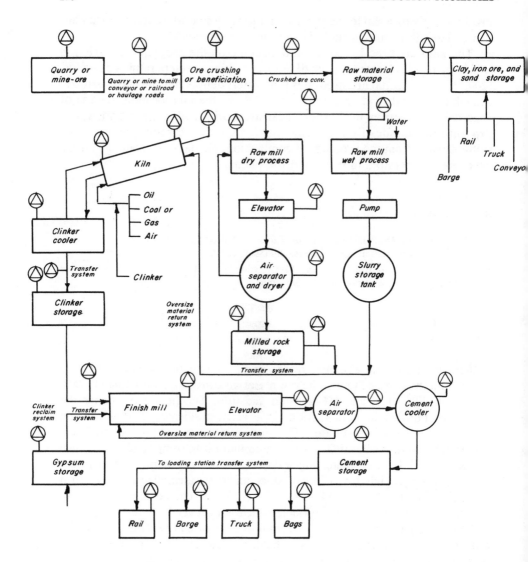

Potential source of dust emissions and/or gases

Figure 3.11-1 Flow diagram for portland cement manufacturing. (From Reference 1, courtesy of American Conference of Governmental Industrial Hygienists)

properly exhausted and the exhaust stream provided with effective air cleaning. The emissions from the kiln are usually handled by electrostatic precipitators and fabric collectors.

The operation of dryers and the calcining kiln results in a potential heat stress problem, and radiant heat is a principal source of the heat load. Radiation shielding of the operators' work locations can assist in reducing this stress. The operation of crushers, grinders, and the rotary kiln also present a noise hazard that may require isolation booths for operators. Exposure to carbon monoxide and sulfur dioxide also may occur from the firing of the kiln.

The alkalinity of the cement presents a dermatitis hazard, and a cement eczema has been reported; it is considered to be due to the hexavalent chromium in the finished product. Cement has been labeled an inert dust, but that practice is being reviewed.

REFERENCE

1 "Process Flow Diagrams and Air Pollution Emission Estimates," American Conference of Governmental Industrial Hygienists, Cincinnati, OH, 1973.

3.12 CHLORINE

Chlorine gas is manufactured by the electrolytic decomposition of sodium chloride brine solution in a cell. The diaphragm cell process produces approximately three-quarters of the United States supply. The balance is produced by chloralkali cell plants. In the diaphragm cell an asbestos diaphragm separates the anode and the cathode. The brine decomposes forming chlorine at the anode, and sodium hydroxide and hydrogen at the cathode.

In the chloralkali mercury cell, shown schematically in Figure 3.12-1, mercury flows through a low voltage-high amperage electrolyzer cell and acts as the cathode. The anode is a stationary carbon or metal component of the cell. The electrolyzer is a long steel chamber approximately 1.2 m (4 ft) wide, 12 m (40 ft) long, and .3 m (1 ft) deep pitched so the mercury and brine will flow from the inlet to the outlet end box. The saturated brine flowing between the cell anode and cathode decomposes, releases chlorine gas at the anode, and forms a sodium amalgam with the flowing mercury. The chlorine product stream is processed as shown in Figure 3.12-1. The mercury amalgam flows by gravity to a companion electrolytic cell called the decomposer or denuder. In the decomposer the amalgam is the anode, graphite cathodes are used, and caustic soda is the electrolyte. Hydrogen is released, the sodium released from the amalgam forms caustic soda, and the "denuded" mercury is recirculated by pump to the electrolyzer cell.

The hazards in the diaphragm cell are limited to accidental releases of chlorine gas and the handling of caustic soda. These hazards also exist in the chloralkali cell, but the major health hazard in this process is exposure to mercury vapor resulting from losses of mercury in the hydrogen stream released from the decomposer and the exhaust from the end boxes of the electrolytic cell. The principal control techniques are the removal of the mercury from these two streams by condensers, scrubbing systems, and treated activated charcoal beds(1). A rigorous housekeeping program has been described to reduce fugitive leaks from the mercury cell. In addition to the mercury exposure, there are safety hazards due to the release of hydrogen in the chloralkali cell.

The precautions to be observed in handling chlorine have been outlined by the Chlorine Institute (2). Skin and eye injury from caustic soda are well known

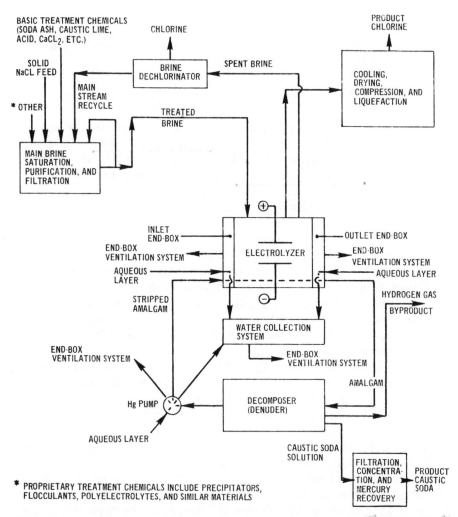

Figure 3.12-1 Flow diagram for chlor-alkali mercury-cell operation. (From Reference 1)

and personnel should wear rubber footwear, aprons and gloves, chemical safety goggles, head protection, and tight fitting sleeves and collar.

Chlorine gas exposures may arise from blow gases from the liquefaction plant, vents from containers during the transfer of chlorine, and from process transfer tanks. Fugitive chlorine leaks also occur from pump and compressor seals, header seals, and air blowing of depleted brine in mercury cell plants (3).

The results of a comprehensive study of mercury-in-air concentrations in chloralkali plants by Smith et al. show a wide range of exposures. The mean exposure for a population of 567 workers was 65 μg/m^3 (4). The authors state that chlorine concentrations in the cell rooms were less than 1.0 ppm and were usually in the range of 0.1–0.3 ppm.

REFERENCES

1 "Control Techniques for Mercury Emissions from Extraction and Chloralkali Plants" U.S. Environmental Protection Agency Publication No. AP-118, 1973.

2 *Chlorine Manual*, The Chlorine Institute, New York, 1969.

3 "Criteria for a Recommended Standard—Occupational Exposure to Chlorine," U.S. Department of Health, Education and Welfare, Publication No. (NIOSH) 76-170, Cincinnati, OH, 1976.

4 R. G. Smith, A. J. Vorwald, L. S. Patil, and T. F. Mooney, *Am. Ind. Hyg. Assoc. J.*, **31**, 687, 1970.

3.13 COTTON

Byssinosis is a disease characterized by shortness of breath, coughing, and a tight chest; this disease occurs in work populations handling certain organic fibers such as cotton, soft hemp, and flax. The symptoms occur on return to the workplace on Monday and gradually subside during the week. The disease is prevalent in the cotton industry and has resulted in a OSHA standard for cotton dust. The etiologic agent for the disease is not known, although it is agreed that cotton plant parts or bract, weed, and fungi, not the cotton fiber itself, are the hosts for the offending agents.

The processing of cotton into textiles involves a large number of specific operations common to many textile plants. An introduction to the textile industry including the purpose of the principal operations and the manner in which textile machines operate is necessary if one is to adequately evaluate exposures and implement appropriate control technology. The processes are described in detail in an industry source volume (1) and in brief form in the NIOSH criteria document (2).

The cotton boll, or fruit of the cotton plant, is now picked almost entirely by machine in the United States. This seed cotton is processed at a local cotton gin to separate the seed from the cotton lint or fiber and remove any trash. After ginning, the lint is cleaned and baled. At this point the biologically active material is included in the cotton, which will be processed in the textile plant.

The quality of the cotton is graded by fiber length and the foreign material present, as shown in Table 3.13-1 (3).

Table 3.13-1   *Cotton Grades*

Cotton Grades	Picker and Card Waste%	Shirley Analyzer Nonlint Control %
Good middling	6.3	1.5
Strict middling	6.4	1.6
Middling	7.1	2.2
Strict low middling	8.2	3.1
Low middling	9.7	4.5
Strict good ordinary	11.2	5.8
Good ordinary	15.0	7.8

Source.   Reference 3.

The yarn manufacturing process carried out in most cotton textile plants is described later in this section. Major attention should be given the carding operation since it is generally acknowledged to be the major source of dust in the plant. In most factories, the dust producing machines are provided with complete enclosures with local exhaust ventilation, air cleaning, and recirculation of the air to the workplace. Since the operations must be conducted at high humidity, it is necessary to add water vapor to the air.

Opening

In this process the baled cotton is opened, the compressed fibers are loosened, and the heavy and bulky impurities are removed to provide material suitable for processing. The dust-producing machinery includes bale breakers, automatic feeders, separators and openers, and mechanical conveyors. Obviously, pneumatic conveying minimizes the dust release.

Picking

A picker includes a feed hopper, beaters, screen sections, and a calendar roll section. This equipment opens and cleans the cotton and forms a continuous web or lap for the card. It is a significant source of dust, and should be provided with an exhausted enclosure.

Carding

As indicated previously, the cards are a major source of dust. The lap formed by the picker is a loosely bound layer of cotton that consists of unopened tufts and tangled fibers. At this point, there is still a large quantity of trash or nonlint material present. The impact of the quality of the cotton on the dust level is shown in Table 3.13-2.

The construction of the card is shown in Figure 3.13-1. The purpose of the card is to remove the trash and short fibers and form a thin, lacy mat with parallel fibers that is then converted to a continuous filament. The lap or mat is

Table 3.13-2 *Dust Levels in Experimental Card Room Produced by Two Cotton Grades*

Sampling Position	Strict Middling Spotted Cotton		Low Middling Spotted Cotton	
	Personal Sampler (mg/m³)	Vertical Elutriator Sampler (mg/m³)	Personal Sampler (mg/m³)	Vertical Elutriator Sampler (mg/m³)
1	1.41	—	4.39	—
2	—	1.20	—	4.12
3	1.57	—	5.39	—
4	1.37	—	4.99	—
5	—	—	—	—
6	—	1.36	—	4.69
7	1.24	—	5.16	—
8	—	1.03	—	4.16
9	1.32	—	4.40	—
10	—	1.25	—	3.87
Sampler average	1.38	1.21	4.87	4.26

Room conditions
    11.5 room air changes per hour
    75°F temperature
    55% relative humidity
    Card production rate—20 lbs/hr
For strict middling spotted cotton
    No significant differences ($p < .05$) among dust levels at different positions or between samplers
For low middling spotted cotton
    Differences are significant ($p < .05$) between extreme sampling position means for both personal and vertical elutriator samplers
Difference is significant ($p < .05$) between dust levels for personal and vertical elutriator samplers

Source. Reference 4.

positioned on a feed roll and fed between the lickerin, a small cylinder covered with wire teeth, and the feed roll. The rotating lickerin "fingers" the lap, opening the tufts of cotton. Opening the tufts releases the trash, which falls out of the lap. The continuous mat is then removed by the rapidly rotating card cylinder that is also covered with wire teeth. Slow moving flats covered with wire cloth pass over the rotary cylinder in a counterdirection, opening and straightening the cotton passing through the interface. Dust and debris are also released at this point.

    The fibrous bed is removed from the card cylinder by the doffer, a small cylinder with needle surfaces. The web is compressed by rolls and fed through a round opening to transform the 100 cm (39 in.) wide web to a round filament or sliver 2.5 cm (1 in.) in diameter, which is coiled in cans. The major sources of

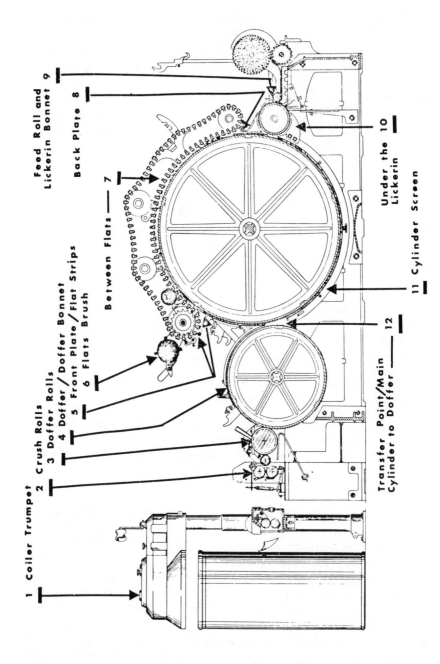

1 Coiler Trumpet
2 Crush Rolls
3 Doffer Rolls
4 Doffer/Doffer Bonnet
5 Front Plate/Flat Strips
6 Flats Brush
7 Between Flats
8 Back Plate
9 Feed Roll and Lickerin Bonnet
10 Under the Lickerin
11 Cylinder Screen
12 Transfer Point/Main Cylinder to Doffer

Figure 3.13-1 Major points of dust generation on card. (From Reference 3)

particulate emissions are shown in Figure 3.13-1. The carding machines are routinely provided with enclosure and suitable exhaust.

Drawing and roving

The drawing process involves feeding several round slivers into a drawing frame where they are compressed into one dense sliver that is coiled again for the next operation. This process pulls the fibers together, arranges them in a parallel orientation, and improves uniformity of the sliver. Although usually enclosed and exhausted, this operation is not a major source of dust.

Roving is a process that again combines many slivers from drawing into one sliver, thus providing more uniformity to the product. The continuous roving is given a twist in the process that improves its strength. It is then wound on a bobbin for spinning. Again, this is not a dusty operation.

Spinning and winding

Spinning reduces the large diameter roving to a small diameter twisted yarn. Since this operation is conducted at high speed, any residual trash in the cotton will be removed and, for this reason, it warrants evaluation. To provide long yarn lengths for continuous weaving operations, yarn is transferred from small bobbins to cones or tubes. This operation is also conducted at high speed and may present a dusty environment. By the time the material is presented for weaving, little trash dust is present and the hazard is minimal.

The NIOSH criteria document identifies the control approaches as: (a) treat the raw cotton to remove or attenuate the offending agent; (b) change the process that is the source of dust; or (c) remove the dust from the air in the workplace (2). The latter approach continues to be the principal control philosophy, although studies continue on the first two alternatives.

The control of dust in the manufacture of yarn cotton has been the subject of a NIOSH-sponsored study (3). The result of this project was the development of design criteria for local exhaust ventilation for processing machinery adequate to maintain airborne concentrations below 0.5 mg/m^3. A variety of exhausted enclosures and air cleaning techniques were proposed in this report. The investigators determined the average air concentrations at the various operations with and without controls, as shown in Table 3.13-3.

Table 3.13-3 *Range of Typical Lint and Dust Concentrations*

Operation	Total Dust (mg/m^3)	Dust Excluding Lint (mg/m^3)
Picking, no control	—	0.6–1.6
Picking, control	0.4–0.7	0.3–0.4
Opening and picking, no control	1.5–9.1	0.2–1.9
Opening and picking, control	—	0.3–0.5
Carding, no control	5.2–21.2	0.3–5.4
Carding, control	0.5–8.4	0.1–4.2

Source. Reference 3.

A study of airborne cotton dust in a large textile plant revealed that present ventilation practice results in the return of most of the respirable dust (5). Recirculation must be curtailed, or improved air cleaning is necessary to control this problem. In addition to local exhaust ventilation, a series of sensible work practices and good housekeeping rules must be followed. The application of single use respirators for protection against cotton dust has been studied by Revoir, and protection factors have been determined during work in cotton mills (6).

A major source of dust in cotton mills is due to blowdown by airlines or blowers. Industry spokesmen feel that deposited lint and dust on machinery and work surfaces represent a major fire hazard and must be removed. In their opinion, vacuuming is inadequate and they recommend judicious use of blowdown with respiratory protection.

REFERENCES

1 D. S. Hamby, Ed., *The American Cotton Handbook,* 3rd ed., Interscience, New York, 1949.

2 "Criteria For a Recommended Standard—Occupational Exposure to Cotton Dust," U.S. Department of Health, Education and Welfare, Publication No. (NIOSH) 75-108, Cincinnati, OH, 1975.

3 H. S. Barr, R. H. Hocutt and J. B. Smith, "Cotton Dust Controls in Yarn Manufacturing," U.S. Department of Health, Education and Welfare, Publication No. (NIOSH) 74-114, Cincinnati, OH, 1974.

4 J. B. Cooke, J. D. Hatcher, and D. L. Smith, *Am. Tex. Rep. Bull.,* (May 1975).

5 Y. Y. Hammad and M. Corn, *Am Ind. Hyg. Assoc. J.,* 32, 662 (1971).

6 W. Revoir, *Am. Ind. Hyg. Assoc. J., 35,* 503, 1974.

3.14 FERTILIZERS

3.14 Natural Fertilizer

The processing of natural fertilizers from excreta of horses, cows, and poultry involves exposure of workers to biological hazards that may cause a variety of infections such as brucellosis, bovine tuberculosis, tularemia, psitticosis, and Q fever. In addition, handling the decomposed material in poorly ventilated areas may involve exposure to hydrogen sulfide, ammonia, and carbon dioxide. The principal controls are the mechanical handling of the product and the provision of adequate dilution ventilation in the workplace. Drying and bagging of the fertilizer may require local exhaust ventilation for dust control.

3.14.2 Mineral Fertilizers (1, 2)

Mineral fertilizers are based on the main soil nutrients, nitrogen, phosphorus, and potassium, and are obtained from the treatment of natural product or produced synthetically.

Normal superphosphate is produced by reacting phosphate rock containing up to 4% fluorides with sulfuric acid to produce a material containing approximately 20% phosphoric anhydride (P_2O_5). Silicon tetrafluoride, carbon dioxide, sulfur dioxide, and particulates are released during this operation, and ventilation controls and air cleaning must be implemented. The material is then cured, ground, and bagged. Gaseous fluorides are released during curing. In one study fluoride concentrations of 1.5 to 3.1 mg/m³ were measured in the storage building where curing takes place (3).

Triple superphosphate containing 40 to 49% P_2O_5 is prepared by the reaction between phosphoric acid and phosphate rock. This operation which may be continuous, involves drying and grinding the product, treating the rock with concentrated phosphoric acid, then mixing and curing the product in large buildings. Silicon tetrafluoride, hydrogen fluoride, ammonia, ammonium chloride, and particulate are the principal air contaminants from these operations.

Since soluble and acidic fluorides can cause skin, eye, and respiratory irritation, adequate personal protective equipment must be made available; showers and eye wash stations must be located in the work area.

Ammonium phosphate is manufactured by combining anhydrous ammonia, phosphoric acid, and sulfuric acid in a reactor vessel. The slightly acidic material is pumped to an ammoniator where additional ammonia is added, causing agglomeration. The agglomerate, diammonium phosphate, is dried, screened, and cooled for packaging. The principal air contaminants are fluorides, ammonia, and particulates.

Ammonium nitrate is manufactured by the neutralization of nitric acid with gaseous ammonia. The product is concentrated to 95% NH_4NO_3 in an evaporator or concentrator, and a solid product called prill is produced by spraying the concentrated product in a tall tower and permitting it to solidify into small pellets during its fall. The product is then dried. The principal hazards to the worker are exposures to ammonia, nitrogen oxides, and nitrate dust.

Potassium-based fertilizers are produced by underground mining of potassium-based minerals including potash (K_2CO_3) and muriate (KCl). The principal exposures are to particulates during the crushing, beneficiation, recrystallization, and drying of the minerals. These materials are frequently mixed with nitrogen and phosphorus-based fertilizers.

Urea-based fertilizers may also be blended with the fertilizers described previously. In one such process, potash, phosphorus rock, and limestone are mixed, crushed and delivered to an ammoniator where the product is coated with a mixture of urea-formaldehyde resin and sulfuric acid. This granular fertilizer is dried, cooled, sized, and bagged. Air contaminants released during this process include formaldehyde, chlorides, and fertilizer dust.

REFERENCES

1 G. H. Collins, *Commercial Fertilizers*, 5th ed., McGraw-Hill, New York, 1955.
2 V. Sauchelli, *Fertilizer Nitrogen*, Van Nostrand Reinhold, New York, 1964.
3 O. M. Derryberry, M. D. Bartholemew, and R. B. L. Fleming, *Arch. Environ. Health*, **6**, 503 (1963).

3.15   FOOD

The health hazards to workers in the food industry include infections resulting from contact with animal products or their waste, exposure to various chemicals used in food preparation and sanitation, and the physical conditions in the workplace, which may vary from high temperature processes including cooking, sterilization, and pasteurization to low temperature facilities such as meat cutting rooms or freezing lockers.

The chief problems for the meat processor are zoonotic disease contracted from diseased animals (e.g., anthrax, brucellosis, glanders, erysipelas, leptospirosis, tularemia, and Q fever) (1). Because of processing demands, the work environment in these plants usually is characterized by high humidity and extreme temperature differences between work areas. Since sanitation is a production requirement, exposure to putrefaction gases such as carbon dioxide and hydrogen sulfide is not a problem. Dermatitis is a common result of contact with such primary irritants as brine and various preservatives. Skin abrasions and cuts on the hand may furnish another mode of entry for pathological organisms when handling meat.

It is common procedure to anesthetize animals prior to slaughter in a room containing high concentrations of carbon dioxide. This may present an asphyxiant hazard to the workers. Forklift trucks powered by internal combustion engines are frequently used for meat transport, and under conditions of poor ventilation in meat lockers, this will result in a serious carbon monoxide exposure. This problem is resolved by improved truck maintenance, use of catalytic converters, improved dilution ventilation, or conversion to electric vehicles.

It is common practice in retail stores in the United States to wrap produce and meat in plasticized polyvinyl chloride film, which can be heat sealed. The film was originally cut by the operator using a hot wire element operating at 250°C (482°F); later a cool rod heated to 135°C (275°F) was introduced after complaints were received of respiratory distress from the fumes and smoke from the hot wire. Although the term "meat wrappers' asthma" was coined to identify the respiratory symptoms, the existence of such a disease entity has not been confirmed. Environmental studies have been conducted at the workplace and in the laboratory to identify the contaminants formed during cutting. Several studies in stores have confirmed the presence of hydrogen chloride at average values of less than 1 ppm and the plasticizer, dioctyladipate, is found in the range of 1 to 10 $\mu g/m^3$. A recent laboratory study identified twelve compounds released during cutting; none of the time-weighted average concentrations exceeded 0.3% of the applicable TLVs (2, 3). A puff of smoke is released each time the film is cut and the instantaneous concentrations of the degradation products may be quite high, causing distress to the sensitized worker (4). Although air contaminants released from both hot wire and cool rod cutting systems are present in air concentrations far below toxic levels, it is a good practice to utilize the cool rod that presents minimal impact on the work environment, or to use a mechanical cutting device, that eliminates all air contamination.

Processing of fish frequently results in malodorous conditions in the workplace. Waste portions of the fish are cooked and pressed to remove oil and water, and the cake is milled and dried in a rotary drier. During cooking of fish, hydrogen sulfide, trimethylamine, and other odorous compounds are released, which may prompt employee and neighborhood complaints.

Milling operations include grinding of various cereals and vegetables to produce fine flour. The major hazard is fire and explosion during grinding, conveying, collection, drying, and storage of the organic material. A series of respiratory problems including bronchial asthma, buccopharyngeal disorders, flour allergy, and chronic rhinitis are attributed to such dust exposures. Exposure to parasites in ground beans and certain cereals causes pruritis and papular skin lesions; some workers are allergic to certain flour molds. One must also consider exposure to insecticide residue in flour. Flour bleaching processes employ a variety of materials including dibenzoyl peroxide, and flour may be enriched by various additives, including nicotinic acid and its amide.

In addition to an organic dust exposure, the baker experiences hot working conditions with highly variable temperatures in the workplace. In the preparation and use of powdered sugar icing, which is frequently mixed and applied by hand, sugar dermatitis occasionally results. Lard oil, when applied to pans by means of swabs, sometimes gives rise to skin irritation.

REFERENCES

1   "Occupational Diseases Acquired from Animals," University of Michigan Education Service, School of Public Health, Ann Arbor, MI, 1964.

2   E. Boettner and G. L. Ball, *Amer. Ind. Hyg. Assoc. J.*, **41**, 513 (1980).

3   W. A. Cook, *Am. Ind. Hyg. Assoc. J.*, **41**, 508 (1980).

4   T. J. Smith, personal communication, 1980.

3.16   GARAGES

Automotive garage work involves a range of metal working operations including sheet metal fabrication, welding, painting, metal cleaning, and abrasive cleaning. These operations are discussed separately in Chapter 2.

The major hazard in service garages is the carbon monoxide emitted by internal combustion engines during the running of vehicles in no-load and dynamometer servicing modes. Tuning-up operations on diesel-powered equipment results in exposure to particulates, oxides of nitrogen, and aldehydes in addition to carbon monoxide. Control may be achieved by opening large service doors to achieve natural ventilation control, although the effectiveness of this measure varies and during cold weather it is impractical. The second control is the use of a flexible exhaust hose discharging outdoors. Leakage around the connection between the tail pipe and the hose and back pressure due to the resistance of the hose make the value of this control questionable. The best ventilation practice is a mechanical exhaust system with the exhaust volume varying from 0.05 to 0.20 m³/s (100–400 cfm), depending on the vehicle horsepower (1, 2).

Dilution ventilation is needed also to eliminate exhaust air contaminants due to vehicle movement in the garage. The ventilation standards for this problem are not as well developed; however, it has been recommended that a minimum of 0.20 m³/s (400 cfm) per work station be exhausted.

Average air concentrations of carbon monoxide are difficult to establish in service gargages due to varying operations. A convenient method to evaluate exposure is the measurement of exhaled breath concentrations before and after work, which allows one to establish approximate carboxyhemoglobin level shifts. Blood carboxyhemoglobin levels of personnel in 35 garages in the United Kingdom increased in 44% of nonsmokers and 20% of smokers (3). In our own experience, it is commonplace to find service garage personnel exhibiting elevated carboxyhemoglobin levels.

An additional problem commonly noted in repair garages that do body work is the lead exposure from filling dents or cracks. Exposures occur during application of the lead-solder filler and during the finishing operations with a disc sander. Hand rasping or scraping does not produce a hazardous amount of respirable dust, but it does present a housekeeping problem. Extensive body work with lead should be done in a properly designed, exhausted enclosure. If exposure is occasional and ventilation is not available, a suitable respirator should be worn. Polyester and epoxy resin systems are also in common use for body repair. Controls should be established to minimize skin contact with the hardeners used in these systems.

The use of gasoline as a solvent is hazardous and should be avoided. Proper degreasing techniques using solvents with high flash points (described in Chapter 2) should be adopted. Workers sandblasting spark plugs for prolonged periods should be checked for possible silica dust exposures. Synthetic abrasives are available for this operation.

With the gradual elimination of tetraethyl lead in gasoline, there is an increase in the aromatic hydrocarbon content of this fuel. Several recent studies have been made of the exposure of garage attendants to benzene while pumping gasoline. The latest study involving a range of types and sizes of service stations revealed that peak concentration may exceed 1 ppm; the time-weighted average concentrations ranged from below the detection level to a high of 0.36 ppm with an average of 0.10 ppm (4). The use of vapor recover systems reduced the air concentration of benzene by a factor of 4.

Personnel who repair brakes are routinely exposed to asbestos particulates abraded from the brake linings and deposited on the surfaces of the brake assembly. It is common practice for garage mechanics to use a high pressure air line to blow out the debris including asbestos dust. Air concentrations during active work may exceed 2 fibers/ml. This practice should be prohibited. Special vacuum cleaner systems with high-efficiency particulate filters are now available for this job.

A noise hazard may exist in service garages as a result of various operations utilizing pneumatic tools such as wrenches for mounting wheels.

REFERENCES

1   Committee on Industrial Ventilation, American Conference of Governmental Industrial Hygienists, *Industrial Ventilation: A Manual of Recommended Practice*, 16th ed., ACGIH, Lansing, MI, 1980.

2   "Garage Ventilation," *Mich. Occup. Health,* **8**, 2(Winter 1962–1963).

3   G. R. Kelman and J. T. Davies, *Br. J. Ind. Med.,* **36**, 238 (1979).

4   R. W. Hartle, "Occupational Exposure to Benzene at Automotive Service Stations." Presented at the American Industrial Hygiene Conference, Houston, TX, 1980.

3.17   GLASS (1)

Nearly every element of this periodic table has been utilized in modern glass technology, and the hazards associated with each have been experienced. The major ingredient of all glass, however, is still silica sand. The common glasses such as soda-lime-silica glass, lead-potash-silica glass, and borosilicate glass contain, in addition to silica sand, the following major constituents: limestone, soda ash, salt cake, lead oxide, boric acid, and crushed glass. Minor constituents include arsenic, antimony, fluoride salts, salts of chromium, cobalt, cadmium, selenium, and nickel, fluorides, sodium silica fluoride, and rare earths. Soda-lime glass represents 90% of the glass produced in the United States.

Since the major component of each batch of glass is sand, this material would seem to present a potentially serious silicosis hazard. In most cases, however, washed sand is used, with a substantial portion of the fine particles removed. It is common to find that airborne dust from the mixed batch contains only from 1% to 5% crystalline silica. Although silicosis is rare in modern glass plants, the methods of handling certain types of sand can still present a dust hazard. The manual unloading of dry sand from boxcars may produce dangerous quantities of fine dust. When wet washed sand is obtained in hopper cars and unloaded by gravity or pneumatically in a totally enclosed system with exhaust ventilation, exposure is minimal. In modern plants, preblended batch materials are obtained in hopper cars, thereby eliminating dust exposure from in-plant handling and mixing.

In the manufacture of optical glass and certain decorative glasses, lead is an important source of employee exposure. Handling of this material, usually in the form of lead oxide, requires personal hygiene procedures and the use of local exhaust ventilation. The other major constituents of glass do not normally present a health hazard, although dermatitis may occur.

The minor constituents have caused health effects, with arsenic the principal offender. Perforation of the nasal septum or severe skin effects due to exposure to arsenic and to highly alkaline constituents of the batch were common occurrences in the past. Modern methods of handling and ventilation control have eliminated most of this trouble.

3.17.1 Glassmaking

The various types of glass manufactured in the modern glass industry are made by two processes: the pot process or the more modern, and now common, tank method. The pot process now serves mainly for the manufacture of high quality glass, such as optical and mirror glass, and for small quantities of specialty glass. In the past this process has been responsible for the greater portion of the silicosis in the glass industry from refractory dust.

Pot melting of glass necessarily introduces the hazard of hand shoveling and filling of the pots. Optical and specialty glasses frequently contain heavy metals such as lead, barium, and manganese. Close attention must be given to handling these toxic materials during the hand-filling process.

The tank process permits enclosed, continuous feeding of batch ingredients, thereby reducing dust exposure at the batch end of the tank. With the introduction of the more efficient tank melting system, the hazards of the pot melting process are disappearing rapidly.

The refractory blocks and bricks used in the construction of the furnaces and tanks contain free silica. Silica brick contains tridymite as its principal constituent. When evaluating the exposure to such dusts during furnace installation and repair, care should be taken to determine the air concentrations of tridymite and cristobalite as well as quartz. In previous years furnace blocks and parts were cut to fit at the installation site. Now, large, well-ventilated, mechanized shops are used to prefabricate refractory furnace parts, which are shipped to the furnace site for installation. Only occasional cutting should be done at the construction site under present-day methods.

Glass objects may be formed by blowing, pressing, casting, rolling, and drawing and by a float method. After forming, all glass objects must undergo a process of annealing to reduce internal stresses in the formed object. This is accomplished in most cases by introducing the objects into long, continuous, annealing chambers called lehrs. Because of their size and the quantity of heat generated, the furnaces introduce a major heat problem.

After the glass has been formed and annealed, it is frequently finished by labeling, smoothing of rough edges, and so on. Glass grinding is done by wet processes, and abrasives such as silicon chloride are used. Polishing is accomplished by the use of revolving felt pads with rouge (iron oxide) as the polishing agent. Glass dust itself is not toxic, since silica is in the combined or silicate form.

Abrasive blasting is sometimes done in enclosed exhausted cabinets with nonsiliceous abrasive materials. Application of decorative enamels by spraying or silk-screen processes introduces the possibility of exposure to solvent vapors, which must be controlled by exhaust ventilation.

In summation, the hazards in conventional glass manufacture are

associated principally with the handling of bulk materials in specialty or
small-run pot production. Tank processes do not, in general, present
hazardous dust concentrations. The major dust hazards are the quartz sand and
special additives listed in Table 3.17-1.

The concentration of furnaces and the handling of the molten glass present
a heat stress hazard in the industry, primarily because of radiation. This
industry was one of the first to demonstrate the importance of reflective
shielding to reduce heat stress.

The hazard from infrared radiation due to the viewing of incandescent glass
warrants the wearing of eye protection on selected jobs.

Table 3.17-1 *Glass Additives*

Process	Materials
Color	Salts of 　Chromium 　Cobalt 　Cadmium 　Manganese 　Nickel 　Selenium
Remove bubbles	Salts of 　Arsenic 　Antimony
Accelerate melting	Fluorine Calcium fluoride Sodium silicafluoride
Improve optical properties	Rare earth metals Thorium

3.17.2 Glass Fibers (Figure 3.17-1)

Fibrous glass wool products used for insulation and acoustical treatment are
formed by air, steam, or flame blowing or by a centrifugal forming technique.
The binder is applied to the hot product; it is collected on a flat moving bed and
conveyed to a curing oven at 200–315°C (400–600°F) and processed as shown in
Figure 3.17-2.

In the manufacture of glass fibers for textiles, gas or oil-fired regenerative
furnaces melt and refine the batch materials at 1540°C (2800°F). The molten
glass is then fed to a forehearth and through a bushing or orifice plate to extrude
a fiber of a given size. The organic binders or "sizes" applied to the fibers
prevent fiber abrasion and act as bonding agents. The product is then cured or
annealed in a furnace and wound on a spool as a continuous fiber or yarn.

The major exposures in the manufacture of fibrous glass products are
mineral dust from the raw materials during batching and furnace charging and
fibrous glass particulates from forming, curing, and product packaging.

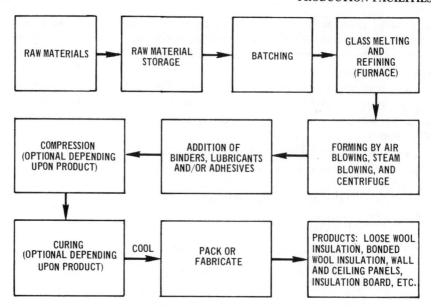

Figure 3.17-1 Typical flow diagram of wool-type glass fiber production process.

Esman et al. have conducted a major survey of particulate fiber concentrations in fibrous glass plants and their data are summarized in Table 3.17-2. Organic vapors and gases may be airborne during curing; their composition depends on the composition of the binder and the spraying technique used to coat the fibers with binder. Fluoride concentrations vary based on the type of flux used in the operation. Combustion products from

Table 3.17-2 *Employee Exposure to Dust and Fibers in Glass Fiber Plant*

Plant	Fiber Diameter, μm	Fiber Product Type	Total Suspended Particulate Matter (mg/m³)		Fibers (fibers/cm³)	
1	1–12	Continuous	0.89	1.12	0.0094	0.25
3	3–6	Loose	0.65	0.46	0.035	0.10
4	1–6	—	1.24	2.26	0.042	0.077
6	5–15	Continuous, loose	0.60	1.04	0.012	0.032
9	7–10	Consumer products	1.33	1.02	0.017	0.014
10	6–16	Continuous consumer products	1.07	0.91	0.002+	0.0032
12	6–115	Continuous loose	0.21	0.16	0.012	0.017
14	6–13	Continuous consumer products	1.42	1.21	0.037	0.031
15	0.05–1.6	Consumer products	0.75	0.67	0.78	2.1
16	6–10	Consumer products	1.07	1.02	0.04	0.12

Source: Reference 2.

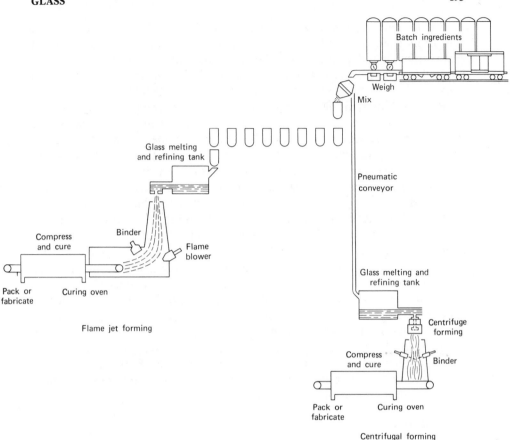

Figure 3.17-2   Schematic of glass forming operations. (Courtesy of P. Reist)

furnaces depend on the fuel in use and may include sulfur dioxide and carbon monoxide.

The glass commonly used for fibrous wool manufacture is a borosilicate type the batch is sand (50%), soda ash (14%), boric oxide (12%), nephaline syanite (12%), flurospar, rutile, and clay. The materials are conveyed pneumatically from storage silos to mixers; minor components are loaded manually to the batch. The mixed material is charged to the furnace. The glass is formed at 1370°C (2500°F) and fed to a refractory tank, or forehearth, where it is held at 1230°C (2250°F). In the flame jet process (Figure 3.17-2) the molten glass flows through a bushing plate located at the bottom of the forehearth and forms a large diameter filament. This primary fiber is heated by a gas flame and the fiber diameter is established by tension on the filament. Once the fiber diameter is established, a phenol-formaldehyde binder is sprayed on the moving fiber bundle and the fibers are collected on a moving conveyer as a blanket. If the product is to be used for home insulation, it is cured at 200–260°C (400–500°C) in an oven; other product lines may not be cured.

In the centrifugal process (Figure 3.17-2), fibers are formed by feeding the glass from the forehearth to centrifugal spinners where the fiber is forced out of small orifices in the spinner head. The fiber is blasted with cold air to attenuate it, the binder is sprayed on the fiber, and the mat of fibers collected on a conveyor belt is cured in an oven at approximately 260°C (500°F).

Pipe insulation, air ducts, and other products are formed from the fibrous mat described earlier. Loose fibrous material may be collected and processed through breakers, blenders, and cards to form a low density web that can be coated with a powdered resin. A mat is formed and the product is cured.

A principal control for particulate emissions at the front end of the process is local exhaust ventilation at all material transfer points and debagging stations. The forming stations also require exhaust ventilation for particulate control and curing ovens must be controlled to minimize exposure to phenol and formaldehyde. In fabrication areas where molded and formed products are manufactured, the particulate exposures must be controlled by ventilation.

The principal exposure in fibrous glass manufacture is to airborne crystalline silica, which occurs during the handling of raw materials at the front end of the process. Studies have shown that much of the airborne dust is respirable. Representative dust concentrations during various operations are not available in the open literature.

Noise levels in the range of 90–100 dBA exist in the furnace area of glass plants originating from high velocity air and gas flow. Furnace rooms and certain product forming areas present serious heat stress conditions frequently exceeding current recommended values.

REFERENCES

1   Section on glassmaking adapted from material written by the late J. Dunn for original Patty volume.

2   N. Esmen, M. Corn, Y. Hammad, D. Whitter, and N. Kotsko, *Am. Ind. Hyg. Assoc. J.*, **40**, 108 (1979).

3.18   IRON AND STEEL

Iron and steel manufacturing presents one of the most diversified sets of occupational health problems of any industry. The modern integrated steel plant includes all operations from the initial handling of coal and ore to the loading of the finished product (Figure 3.18-1). The major air contaminants encountered in the industry are listed in Table 3.18-1. In this section special attention will be given to coke production, blast furnace production of pig iron, and the steelmaking processes (1, 2).

3.18.1   Intermediates and Products

Pig iron, the principal intermediate in steel production, is made from iron ore, limestone, and coke reacted in the blast furnace. This intermediate contains

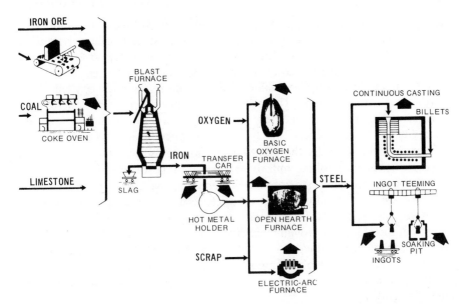

Figure 3.18-1 Flow diagram of an iron and steel plant. (Courtesy of the American Iron and Steel Institute.)

Table 3.18-1 *Major Air Contaminants in Iron and Steel Industry*

Operation	Exposure
Dust	
Mining	Ore and coal dust
Ore sintering and pelletizing	Iron oxide
Coke ovens	Coke oven emissions
Refractory handling	Silica dust
Foundries	Silica sand
Metal fume	
Furnaces	Iron oxide
Scarfing operations	Iron oxide
Scrap preparation	Lead fume
Galvanizing	Flux fume, zinc
Leaded and ferromanganese steels	Lead and manganese fume
Gases and vapors	
Blast furnace	Fluorides, CO
Coking operation	CO, SO_2, H_2S
Welding	Ozone, oxides of nitrogen
Maintenance and cleaning motors	Solvent vapors
Mists	
Pickling	Sulfuric acid mist
Plating	Various
Spray painting	Lead paints spray mist

excess carbon, manganese, phosphorous, sulfur, and silicon. It is not used directly as a product but is converted into cast iron or steel. Cast iron is an alloy that contains iron, 2–4% carbon, silicon, and manganese and is usually produced in a cupola furnace at the steel plant or in foundries.

Steel ingots or castings are made from pig iron in the basic oxygen, open hearth, or electric furnace. The conversion from steel to iron is a refining process in which the carbon content is fixed at a lower percentage and certain other elements are reduced in composition. Specific alloy steels may be fabricated at this time by the addition of alloying metals. The cast ingot is processed through a rolling mill to produce a range of products.

Alloy steels are either steels to which elements other than carbon are added or carbon steels containing greater than 1.65% manganese, 0.60% silicon, or 0.60% copper. These steels frequently have alloying materials that permit the alloy steel to achieve improved mechanical properties by heat treatment. The elements frequently added to carbon steel to form alloy steels include vanadium, chromium, molybdenum, aluminum, and nickel. Other materials including cobalt, columbium, and tungsten are found in special steel alloys made for cutting tools.

3.18.2 Manufacture of Coke (Figure 3.18-2)

Bituminous coal is crushed, cleaned, dried and shipped to the coke plant. In by-product coke ovens, the coal is baked in a slot oven in the absence of air at temperatures of approximately 1090°C (2000°F) for 14–20 hr. The oven is maintained under a vacuum, and as the volatile materials are driven off they are collected and recovered at the by-product plant.

Coke oven workers are exposed to a range of particulates, gases, and vapors probably not equaled in any other industrial setting. It is estimated that one quarter of the total weight of coal is evolved as gases and vapors in the coking processes and that over 2000 different chemical species may be formed. The principal air contaminents include CO, CO_2, H_2S, SO_2, NH_3, aromatic hydrocarbons, and polynuclear aromatic hydrocarbons. One ton of coal will produce $0.030\ m^3$ (8 gal) of tar, 9 kg (20 lbs) of ammonium sulfate, $170\ m^3$ (6000 ft)^3 of surplus coal gas, $0.01\ m^3$ (3 gal) of light oil, and 0.056–$0.11\ m^3$ (15–30 gal) of ammonia liquid.

An epidemiological study of the steel industry conducted in the 1960s revealed that coke oven workers employed for more than 5 years had a lung cancer mortality rate 3.5 times that expected, the rate for topside workers was 10 times that expected (3, 4).

The excess lung cancer is believed to be correlated with the concentration of coke oven particulate emissions identified as the benzene soluble component of the airborne particulate and include known carcinogens such as benz(a)pyrene, benz-fluoranthene and chrysene. Several studies of air concentrations have been presented in the literature.

The coke plant shown in Figure 3.18-2 consists of a stockpile of crushed

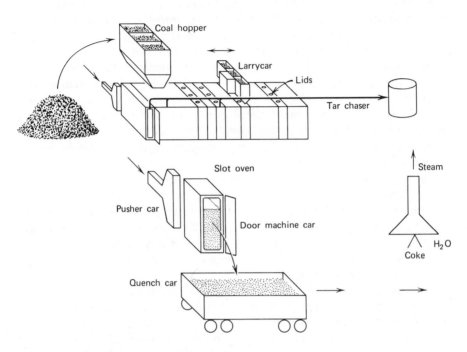

Figure 3.18-2   Schematic of coke plant.

bituminous coal, a battery of coke ovens, and a by-product plant. The coke battery consists of a number of slot ovens that are 8–15 m (25–50 ft) long, 3–8 m (10–25 ft) high, and 35–60 cm (14–24 in.) wide. These slot ovens alternate with spaces heated by the burning of coke gas produced during the operation. A series of round charging ports provided with lids in the top of the oven permit the charging of 15–20 tons of coal to one slot oven. The ends of the ovens are large refractory doors.

In evaluating worker exposure on coke ovens, one must be acquainted with the various jobs. Coal is dumped from a storage bunker to the hopper of a larry car that runs on rails on the top of the battery. The larry car operator moves the car to a given oven. A lidsman manually removes the lid from that oven, the car is positioned, and the coal is dumped to the oven. At this time there may be a violent eruption of dust and some combustion with exposure of both the larry car operator and lidsman to a heavy contaminant cloud. The larry car returns to the bunker station for another load of coal. The lidsman sweeps up the loose coal, replaces the lids, and seals the lids with a luting (clay-water) mixture. This activity is repeated many times during the shift. In some modern batteries the coal is transferred by pipeline directly to the oven.

The coal is baked at 1090°C (2000°F) for 16–20 hr. During this period off-gases are collected by a vacuum main and recovered in the by-product plant. At the end of the coking period, the ends of the slot oven are opened and the pusher car operator positions that equipment at the oven. A ram mounted on the car pushes the red hot coke out of the oven into a quench car. The coke is moved to the quench tower by the quench car operator, and the coke is deluged with water. After quenching, it is dumped for conveyor transport.

In addition to the larry car operator and the lidsmen, the other topside workers include a luterman, gooseneck cleaner, and various maintenance personnel. The so-called sidemen include the operators of the pusher car and the quench car and door maintenance personnel. The coke oven worker is exposed to coal dust, particulate coke oven emissions with a significant benzene-soluble component, carbon monoxide, hydrogen sulfide, sulfur dioxide, and a range of other contaminants in low concentrations. Heat stress is also a significant physical hazard on coke ovens.

Exposures in by-product plants include carbon monoxide, ammonia, benzene, carbon disulfide, and other contaminants. Since the operations are conducted in an enclosed system, the principal difficulty is unexpected leakage, resulting in high concentrations for brief periods. Plant maintenance, intelligent supervision, and thorough training are required to ensure safety with respect to these exposures.

A recent NIOSH publication presents the available control techniques for the coke oven environment (5). The proposed priority list for controls include side door leaks, topside leaks, charging emissions, pushing emissions, and coal shoveling practices. The principal engineering controls include staged charging to reduce charging emissions, pipeline charging that would eliminate the topside jobs of larry car operators and lidsmen, mechanical lid lifters for larry cars, and dry rather than water quench of the coke. Provision of jumper lines that place the charging oven under negative pressure has demonstrated value. The use of air conditioned enclosed stand-by pulpits, and restrooms with a filtered air supply can significantly reduce exposure. Fugitive leaks from lids, standpipes, goosenecks, and doors can be reduced by proper maintenance and luting. Installation of sheds over the ovens as a means of environment pollutant control increases worker exposure and does not appear to be an appropriate control.

3.18.3 Blast Furnace (Figure 3.18-3)

Taconite, a common iron ore in the United States, contains 23 to 25% iron. The ore is crushed and beneficiated by either magnetic or flotation methods and is transported to the steel mill as pelletized material containing 60–70% iron. Direct reduction techniques have now been introduced to provide material with an iron content in excess of 90%, which may permit by-passing the blast furnace.

In the handling of the ore, many "fines" or small particles are generated. This material, with recovered blast furnace dust, is mixed with coal dust or "breeze," spread on a traveling grate, and "fixed" in a sintering furnace. After sintering, this porous cake is crushed and is used as a part of the blast furnace

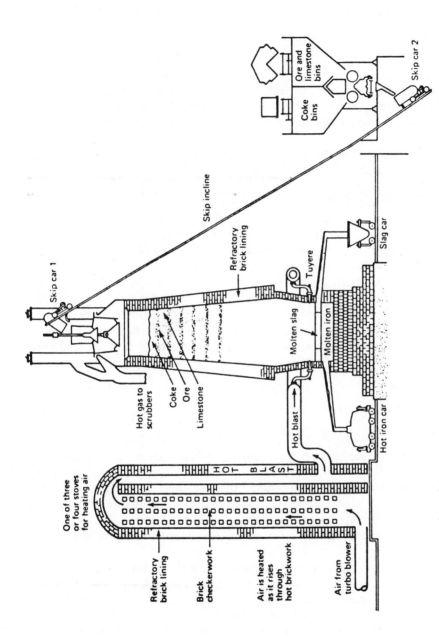

Figure 3.18-3 Blast furnace. (Courtesy of American Iron and Steel Institute.)

Skip car 1

Skip car 2

Ore and limestone bins

Coke bins

Skip incline

Refractory brick lining

Hot gas to scrubbers

Coke

Ore

Limestone

Tuyere

Molten slag

Molten iron

Slag car

Hot blast

Hot iron car

One of three or four stoves for heating air

HOT BLAST

Refractory brick lining

Brick checkerwork

Air is heated as it rises through hot brickwork

Air from turbo blower

charge. Major dust hazards may occur in sintering operations, and these require control by local exhaust ventilation.

The iron ore and limestone are stockpiled, weighed, and loaded into skip cars at the stockhouse for the charging of the blast furnace, as shown in Figure 3.18-3. Charging of coke frequently is done automatically. The raw materials are charged to the blast furnace, a cylindric tower lined with refractory brick, through a charging "lock" consisting of small and large bells. The raw materials increase in temperature as they sink down into the stack of the furnace. Oxygen is removed from the ore in the top section of the furnace, and midway down the flux limestone reacts with impurities to form a slag that absorbs the ash from the coke.

The coke, reaching the base of the stack, burns causing smelting reactions that release iron from the ore. Carbon monoxide formed during burning of the coke is a reducing gas that flows upward, burning and reducing the iron oxide to iron. At the bottom of the furnace there is a pool of molten iron 1.2-1.5 m (4–5 ft) deep with the slag floating on top. The slag is drained off through the slag notch to a ladle, and the molten iron is removed at the tap hole to a hot metal car for transport to steelmaking facilities such as the open hearth or basic oxygen furnace. The vent gases from the blast furnace are used to preheat the blast furnace combustion air.

The blast furnaces present a significant exposure to carbon monoxide, especially during maintenance operations, and several deaths have occurred in the industry from this exposure. Concentrations of carbon monoxide should be monitored before maintenance operations commence, and self-contained breathing apparatuses must be available for escape purposes.

3.18.4 Steelmaking

The common steelmaking facilities in the United States are the open hearth furnace, the electric furnace, and the basic oxygen furnace (BOF). The open hearth furnace, shown in Figure 3.18-4, is being replaced rapidly by the BOF. In the open hearth, a mixture of scrap iron, steel, and pig iron is charged to a shallow hearth or basin and exposed to open flame from the combustion of oil, coke oven gas, or natural gas. Limestone is used as a flux and slag is removed; oxygen is sometimes injected in the melt. The steel can be tapped off after 5 to 8 hr.

The electric furnace (Figure 3.18-5) is used to produce steel alloys and stainless and specialty steels, although it is increasingly used to make conventional carbon steel. In this furnace three carbon electrodes are positioned in the cavity of the furnace over the charge. The current arcs from one electrode to the charge and then from the charge to the next electrode, providing an intense heat source and melting the solid charge. Limestone and

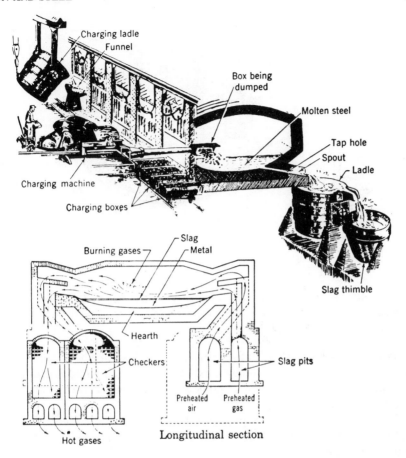

Figure 3.18-4 Open hearth furnace. (From *Foundry Engineering* by H. F. Taylor, M. C. Fleming, and J. Wulff, Copyright 1965. Used with the permission of John Wiley and Sons, Inc.)

flux are introduced to remove impurities to a slag layer. Carbon is removed by direct oxygen injection.

The principal steel producing furnace in the United States is the BOF. The furnace, shown in Figure 3.18-6, is a refractory lined steel shell supported in horizontal trunnions. The furnace is tilted for charging with steel scrap from a scrap charging car, molten pig iron is then introduced into the top of the furnace, and finally a water cooled oxygen lance is lowered into position 1.8 m (6 ft) above the metal surface. After initial heating, lime and fluorspar are added by chute. The oxygen combines with carbon and impurities, and these impurities float off as slag under the action of lime and fluorspar. Alloying

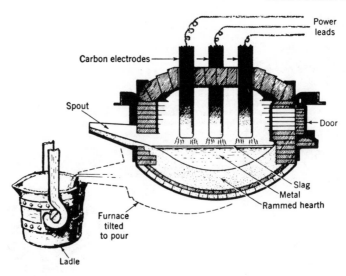

Figure 3.18-5 Electric furnace. (From *Foundry Engineering* by H. F. Taylor, M. C. Fleming, and J. Wulff, Copyright 1965. Used with the permission of John Wiley and sons, Inc.)

metals can be added directly to the heat by chute. A smoke hood is positioned at the throat of the furnace to remove metal fume. The anticipated problems of exposure to metal fume, carbon monoxide, heat, and noise are encountered in BOF opeiations.

Several different steels are produced by these furnaces. The basic carbon steel is produced as described, and the alloying materials are then added to meet certain specifications.

The molten steel produced in the above-mentioned furnaces is translated into steel products in myriad ways (Figure 3.18-7). The molten steel in the ladle may be poured or "teemed" into cast iron ingot molds. The molds are tapered to facilitate removal of the solid steel. After stripping the mold from the ingot, the ingot is placed in soaking pits or holding furnaces where it is held until a uniform temperature is reached throughout the ingot. The ingot is then removed from the soaking pit to mills to form blooms, slabs, and billets. Blooms have a square cross section, slabs are rectangular, and billets are long with small cross sections.

In some facilities the molten metal is poured directly in a semifinished form by-passing teeming, stripping, soaking, and rolling. The intermediate processing of steel includes hot and cold rolled mill products, tin mill products, hot dip galvanizing, structural shapes, and product manufacture of steel rod and wire, pipe and tubing.

A potential lead exposure exists during the production of leaded steel, but control by local exhaust ventilation is possible. Special steels may contain

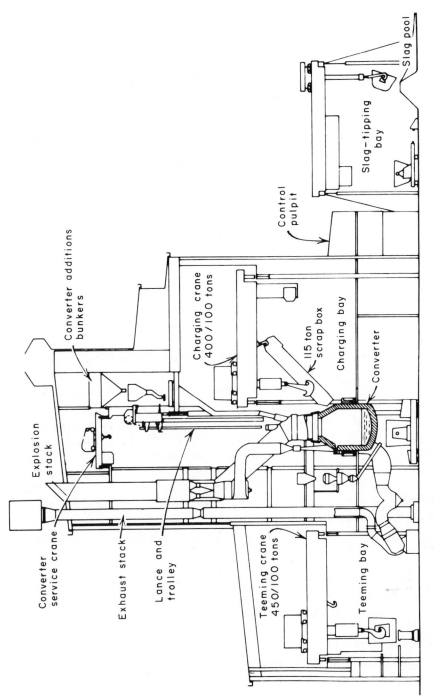

Figure 3.18-6 Basic oxygen furnace (BOF). (From Engineering Design Problems of Large Iron and Steel Making Furnaces. ISI Publication 136. Copyright 1970. Used with the permission of the Metals Society. London.)

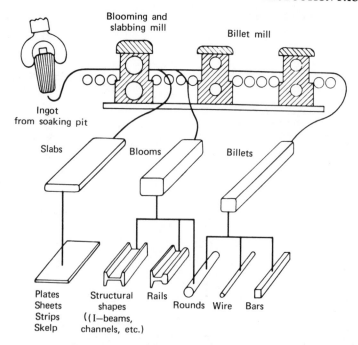

Figure 3.18-7 Steel products. (Courtesy of the American Iron and Steel Institute.)

nickel, bismuth, chromium, ferromanganese, tungsten, and molybdenum. Fluorides may be encountered in connection with certain iron ores.

The production of tin plate involves exposure to acids, and terneplate also involves lead. Zinc baths for galvanized steel have been found to be contaminated with lead in amounts sufficient to constitute a hazard. Iron and steel plants also have plating, spray painting, and welding operations with the attendant hazards noted in Chapter 2. Metal fume exposures occur during oxygen injection at blast furnaces and in electric arc furnaces. These emissions require ventilation control and air cleaning (6, 7).

The major silica hazard occurs during the installation of refractory brick materials containing high concentrations of quartz in the lining of furnaces and ovens.

Heat stress is a problem in coke oven operations, basic steelmaking, and final mill operations. The principal heat load is due to radiation from furnaces and the molten metal. The industry has utilized the following controls:

1 Protective shields for radiation.
2 Air conditioning of control stations, pulpits, and crane cabs.
3 Spot cooling of work sites.
4 Personal protective clothing such as aluminized garments equipped with vortex coolers.

The noise hazards in this industry are pervasive and necessitate a variety of controls, including modifications of equipment, mufflers for air exhaust and intakes, isolation and enclosures, and personal protective devices.

REFERENCES

1 "The Making of Steel," American Iron and Steel Institute, Washington, D.C., undated.
2 "Steel Processing Flow Charts," American Iron and Steel Institute, Washington, D.C., undated.
3 J. W. Lloyd, *J. Occup. Med.,* **13**, 53 (1971).
4 C. K. Redmond, A. Ciocco, J. W. Lloyd, and H. W. Rush, *J. Occup. Med.,* **14**, 621 (1972).
5 J. W. Sheehy, "Control Technology for Worker Exposure to Coke Oven Emissions," U.S. Department of Health, Education and Welfare, Publication No. (NIOSH) 80-114, Cincinnati, OH, 1980.
6 American Iron and Steel Institute, *Steel Mill Ventilation,* AISI, Washington, D.C., 1965.
7 "Criteria for a Recommended Standard—Occupational Exposure to Coke Oven Emissions," U.S. Department of Health, Education and Welfare, Publication No. (NIOSH) HSM 73-11016, Cincinnati, OH, 1973.

3.19 LEATHER

Anthrax is a much cited biological hazard in the leather industry. This was a serious problem in the United States, especially from skins imported from anthrax-infested areas, but regulations and improved methods of treating imported hides have done much to bring the exposure under control. Anthrax is usually acquired by contamination of wounds or abrasions by *Bacillus anthracis,* but the organism may be inhaled or ingested. Workers handling raw hides or skins should be included in a medical control program.

The various processes in leather processing are shown in Table 3.19-1. Dried animal hides received for processing have been treated with insecticides and possibly other materials such as salt to stabilize the hide and prevent decay. At the tanning plant, the hide is first treated with a disinfectant and a surfactant in a soak tank (soaking). Excess flesh is removed from the inside of the skin mechanically (fleshing), and the hides are then "limed" in milk of lime to loosen the epidermis and remove soluble protein and fat (unhairing). Various chemicals including sodium sulfide, sodium hydrogen sulfide, and dimethylamine may be added during "liming" to enhance the unhairing. Deliming is carried out in a solution of sulfurous acid or ammonium salts. Enzymes are added to remove undesirable constituents from the skin (bating). The hides are then pickled in solutions of sulfuric and hydrochloric acids to prepare the skin for tanning and treating with 2-naphthol and *p*-nitrophenyl to prevent molding of the skin. These procedures are normally carried out in series in a large wooden tumbling barrel (1). A potential hazardous exposure to hydrogen sulfide results from the reaction of acid with residual sulfide solutions on the hides due to poor washing. Several deaths from hydrogen sulfide in leather plants have been described in European literature.

Table 3.19-1 *Operations in Leather Processing*

First stage (wet operations)	Second stage (dry operations)
Trimming and sorting	Drying
Soaking	Conditioning
Fleshing	Staking
Unhairing	Buffing
Bating	Finishing
Pickling	Plating
Tanning	Measuring
Wringing	Grading
Retanning, coloring	
Setting out	

The tanning process also is conducted in a tumbling barrel using chromic acid, alkalies such as trisodium phosphate and borax, oxalic acid, formaldehyde, and natural and synthetic vegetable tanning materials. In addition to exposure to these chemicals, the putrefaction of animal material may cause exposure to carbon monoxide, hydrogen sulfide, and methane.

The tanned hides then go through a series of finishing operations: pressing, splitting, shaving, sanding, buffing, finishing, waxing, oiling, and so on. Sanding and buffing result in an exposure to leather dust; local exhaust ventilation must be provided for such operations. Certain machines may also cause a serious noise hazard.

To prevent the generation of carbon monoxide, the tanning solutions should be changed frequently. Hydrogen sulfide may be present as a product of putrefaction or as a result of the treatment with acid of hides containing sulfides. Good housekeeping must be stressed, and workers should be urged to minimize skin contact with hides and chemicals. Protective boots, aprons, and gloves should be worn, and adequate shower facilities must be available. All operators should be included in a periodic medical surveillance program. In the finishing plant a variety of surface coatings with associated organic solvents are encountered. The spray application of the finish and its drying must be performed with good local exhaust ventilation. A rigorous program for entry to confined spaces such as tanks, pits, sewers, and unventilated basements must be promulgated and observed by all plant personnel, especially maintenance crafts persons.

REFERENCES

1 New England Tanners Club, P.O. Box 371, Peabody, MA, 1973.

3.20   LIME

The lime products listed in Table 3.20-1, are used in masonry mortars, wall plaster, soil stabilizers, as fluxes in steelmaking, as glass desiccants and in water treatment. Lime is employed in the manufacture of insecticides and bleaches also. Calcium oxide or quicklime (CaO) is produced by calcining limestone ($CaCO_3$) in a vertical or rotary kiln. Water is added to quicklime in hydrators to form hydrated or slaked lime, $Ca(OH)_2$. This material suspended in water is known as milk of lime.

The reactions based on high calcium limestone are

$$CaCO_3 + heat \rightarrow CaO + CO_2$$
$$CaO + H_2O \rightarrow Ca(OH)_2 + heat$$

If dolomitic limestone is the starting material, the reactions are

$$CaCO_3 + MgCO_3 + heat \rightarrow CaO \cdot MgO + 2CO_2$$
$$CaO \cdot MgO + H_2O \rightarrow Ca(OH)_2 \cdot MgO + heat$$

Protection of the eyes and of the skin presents the major environmental control problem in the production of calcium oxide or quicklime. The material is of small particle size and is very irritating to mucous membranes and moist skin. It combines with water with the evolution of heat to form calcium hydroxide, which is nearly as caustic as potassium hydroxide.

Air-slaked lime, which is almost 100% calcium carbonate, has mild causticity and usually attacks mucous membranes. If warm, it may cause dermatitis after prolonged exposure. Quicklime rarely affects the lungs, since it is so irritating to the upper respiratory tract that exposure for the necessary period does not occur. It rapidly produces coughing and sneezing, which limit further exposure.

Table 3.20-1   *Types of Limes*

Type	Formula
High-calcium limestone (5% or less $MgCO_3$)	$CaCO_3$
High-calcium quicklime	CaO
High-calcium hydrate	$Ca(OH)_2$
Dolomitic limestone (over 35% $MgCO_3$)	$CaCO_3 \cdot MgCO_3$
Dolomitic quicklime	$CaO \cdot MgO$
Dolomitic monohydrate, type N	$Ca(OH)_2 \cdot MgO$
Dolomitic doublehydrate, type S	$Ca(OH)_2 \cdot Mg(OH)_2$

Workmen at lime kilns may be exposed to dangerous concentrations of carbon monoxide, carbon dioxide, hydrogen sulfide, and arsine. Dust exposures occur during all handling and transfer operations. The principal controls for particulates are local exhaust ventilation at lime hydration operations, conveying transfer points, and pulverizing and bagging stations. Eye and hand protection may also be required. As in all calcining operations, heat stress due principally to radiation is a potential problem and must be handled by conventional controls.

3.21 PAINT

As discussed in Chapter 2, paint, in its simplest form, is a pigment suspended in a vehicle that may be an oil, a varnish, or a natural or synthetic resin. The batch manufacturing process for paint shown in Figure 3.21-1 includes paste mixing,

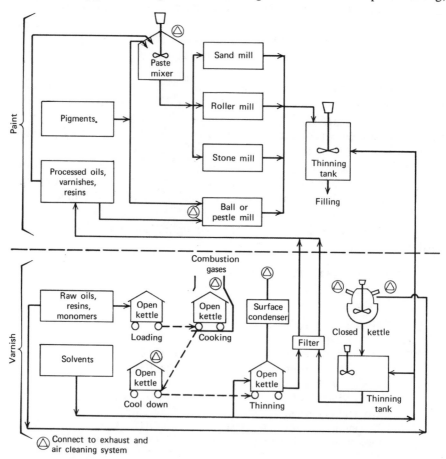

Figure 3.21-1 Flow diagram for paint and varnish operations. (From Reference 1, courtesy of American Conference of Governmental Industrial Hygienists)

dispersion, and thinning (1). The pigments are first mixed with the resin vehicle in a low speed paste mixer and then gravity fed to a mill for dispersion of the pigment. The dispersed vehicle-pigment mix is then transferred to a closed tank for thinning and dispensing into containers.

The production of the binders for lacquers, spirit varnish, and epoxy systems may require only cold blending of natural or synthetic resins in a solvent. Most conventional paints, however, require heating combinations of film-forming materials in reaction kettles, as shown in Figure 3.21-1. The process shown involves open kettles of 0.4-1.5 m³ (100–400 gal) capacity, a technique encountered in older plants. Modern kettles may utilize large vessels equipped for close temperature control, closed transport of materials, mixing, and inerting when handling flammable thinners. The oils, resins, or monomers, that are the basis for the paint resin in a given system, are transferred manually or by pump to the kettle, which is transferred on a dolly to an exhausted heating station. After cooking at 340°C (650°F), the batch is taken to another plant site for thinning to the desired percentage solids. The product is filtered and is used in the paint manufacturing process.

Workers in paint factories are exposed to mineral dust fillers, pigments, and volatile thinners. The pigment and filler dust exposures occur during charging of the paste mixer and mills. Ventilation control may be achieved with local hood exhaust at charging ports with face velocities of 0.5–1.0 m/s (100–200 fpm) (1). Cooking must be conducted in enclosed hoods with a minimum exhaust of 0.15–0.20 m³/s (300–400 cfm) per kettle to control releases, which may include acrolein and other aldehydes, phthalic anhydride, mercaptans, phenols, acrylics, and organic acids. For a 8m³ (2000 gal) closed kettle, one must exhaust 0.005–0.01 m³/s (10–20 cfm) of the inert gas plus 0.01–0.02 m³/s (20–40 cfm) for the reaction gases.

Mineral fillers are used to produce low luster paints and enamels. These mineral fillers are often made of crystaline quartz, therefore, they warrant close control during debagging and charging to mix vessels. Good housekeeping and personal cleanliness must be stressed in paint manufacturing to minimize exposure and prevent dermatitis.

No data have been published in the United States on exposure during paint manufacture; however, Ulfvarson has presented data in a comprehensive survey of 10 paint factories in Sweden that employ 40% of the employees in that country's paint industry (2). Samples were collected by personal air sampling methods where possible. The particulate exposures included quartz, fibrous dust, hexavalent chromium, and lead. As shown in Table 3.21-1, hazardous concentrations were encountered in a number of locations.

Solvent concentrations were also evaluated at these plants. The air concentrations in Table 3.21-2 are expressed in terms of the Swedish occupational health standards. That is, if the mean value noted is 2.0, then the mean concentration was two times the health standard. Where mixed solvents were encountered, the ACGIH technique for evaluating mixed exposure was used in the calculation utilizing short term exposure values of 1.25 to 1.5.

Again, these data indicate that significant exposures to solvents occur in the Swedish paint industry. It is reasonable to assume that U.S. facilities have comparable exposures.

Table 3.21-1 *Summary of Observed Dust Concentrations*

Materials[a] Charged	Number of Samples	Dust Concentration (mg/m^3)	Quartz (mg/m^3)	Fibers/ cm^3	CrO$_3$ (mg/m^3)	Pb (mg/m^3)
			Concentration of Components in Selected Samples (number of such samples in parentheses)			
Inorganic	3	2.8–5.6		0.31(1)[a]	0.005(1)	
Inorganic	4	4–21	0.02(1)			
Organic	5	2.1–14				
Inorganic	5	10–28	0.08(1)		0.06–1.6	
Inorganic + organic	9	3.4–41	0.04(1)		0.04–0.44(2)	0.06(1)
Inorganic + organic	11	1.7–28	0.9(1)		0.2–1.5(2)	0.49–4(3)
Inorganic + organic	9	2–25	0.01(1)	1.2–1.3(2)	0.12(1)	
Inorganic	5	3.7–14		1.8–2.3(2)		
Inorganic	6	1.9–40		5(1)	0.01–0.03(3)	0.006–0.007(3)
+ organic						
Inorganic + organic	4	7.7–70			0.003(3)	

Source: Reference 2.
[a]Pigments, extenders, and so on.

Table 3.21-2 *Summary of Solvent Concentrations*

Operation	Number of Observations	Mean and Range of Sums of Standardized Concentrations[a]	Principal Solvents (number of cases in which the solvent was found)
Charging solvents	33	mean = 2.0 0.2–16	Xylene (16), mesitylene (4), toluene (4), styrene (2), butanol (9), esters, and others
Pigment dispersion (grinding, roll-milling, etc.) + emptying of vessels	18	mean = 1.5 0.2–4.4	Xylene (13), butanol (4), and others
Tinting, thinning	14	mean = 0.9 0.1–2.0	Xylene (11), butanol (3), and others
Can filling, paints	39	mean = 1.3 0.02–6.6	Xylene (23), alkanes (4), butanol (7), benzene (4), toluene (6), and others

Table 3.21-2 *Summary of Solvent Concentrations—Continued*

Operation	Number of Observations	Mean and Range of Sums of Standardized Concentrations[a]	Principal Solvents (number of cases in which the solvent was found)
Can filling, thinners	14	mean = 1.8 0.1–7.4	Toluene (3), xylene (5), trichloroethylene (3), esters (2), acetone (1), and others
Manual cleaning of equipment with solvents	51	mean = 5.7 0.5–30	Xylene (33), butanol (8), toluene (13), methylene chloride (9), esters (7), and ketones (4)
Laboratory work cleaning with solvents, spraying paints, grinding, oven drying	22	mean = 0.9 0.06–3.0	Xylene (11), toluene (7), ethyl acetate (4), and others

Source: Reference 2.

a. Standardized Concentration = $\dfrac{\text{Observed Concentration}}{\text{Health Standard}}$

REFERENCES

1 American Conference of Governmental Industrial Hygienists, "Process Flow Diagrams and Air Pollution Emission Estimates," ACGIH, Cincinnati, OH, 1973.
2 U. Ulfvarson, G. Rosén, M. Cardfelt, and U. Ekholm Kemiska Risker i Färgindustrin (Chemical Hazards in the Paint Industry) (in Swedish), Uppdragsrapport Dnr 4979/75, Arbetarskyddsstyrelsen, Arbetsmedicinska avdelningen, Stockholm, 1976.

3.22 PETROLEUM

3.22.1 Exploration and Transportation

The occupational health problems encountered in exploration and extraction are associated with both the severe environmental conditions under which drilling is conducted and the nature of the oil (1, 2). Chemicals such as formaldehyde and hydrogen chloride may be pumped into the well during drilling. If the crude oil is sour, hydrogen sulfide may be evolved. Other possible crude oil contaminants are vanadium and arsenic.

In the transport and storage of the oil, hydrogen sulfide may be encountered. The principal exposure occurs during the cleaning of shipboard and tank farm storage tanks. Tank cleaning procedures involving direct entry by workers may also constitute a hazard because of oxygen deficiency resulting from previous inerting procedures, rusting, and oxidation of organic coatings. Carbon monoxide may also be present in the inerting gas. In addition, depending on the characteristics of the product previously stored in the tanks, one may encounter hydrogen sulfide, metal carbonyls, arsenic, and tetraethyl lead. Detailed procedures for tank entry including ventilation purging, environmental testing, and respiratory protection must be prepared.

3.22.2 General Refinery Problems

The United States has 285 refineries employing 70,000 workers. These refineries process a variety of petroleum products with low flash points; thus the principal hazards at refineries are those of fire and explosion. All operations are designed to prevent such catastrophes, nevertheless, serious refinery fires occur.

The principal exposures in refinery operations occur during shutdown, maintenance, and startup procedures. So-called plant turnarounds must be carefully scheduled and step-by-step procedures developed to ensure that operations are conducted safely. The exposures vary depending on the unit being maintained. Difficult control problems include abrasive blasting and welding in enclosed spaces. Asbestos insulation is still in use in refineries, and rip-out and repair of this material are short term but potentially hazardous operations.

Because of the many pumps, blowers, burners, fans, flares, and steam and air releases, a significant noise exposure may exist in certain points of the refinery.

3.22.3 Specific Refinery Processes

One must review the basic processes to understand the source of potential health hazards in refineries (3). A modern refinery is a maze of furnaces, heat exchangers, pumps, tanks, fractionating columns, pipes, pipe fittings, and valves. However, the refinery can be divided into a number of unit operations to allow the hygienist to make a reasonable inventory of health hazards.

The initial refinery process, crude distillation, involves separation of the oil into various fractions or "cuts," which have specific boiling temperature ranges, utilizing a distillation tower. These fractions lack a definite chemical formula, being defined solely by their boiling ranges.

Atmospheric distillation towers are cylindrical towers that contain a series of horizontal trays. Heated oil pumped to the tower splits into two phases; the vapor starts to ascend the tower while the unvaporized reduced crude is withdrawn from the bottom of the tower for processing in a vacuum distillation unit. The vapor cools as it flows upward, and the high boiling point fractions condense out in the lower trays.

The trays are designed so that the vapor must bubble through the cooler condensate in the bottom of the tray; the lighter hydrocarbons will boil off and the heavy fractions in the vapor are condensed. The lower boiling point fractions continue upward until all condense on selected trays. The final uncondensed material is taken to a separate condenser. Potential health hazards associated with primary distillation include hydrogen sulfide, hydrocarbon vapors (aryl and alkyl), heat, and noise.

A single distillation does not produce the desired quantities or quality of each product. Therefore, less desirable products must be transformed to more desirable products by splitting, uniting, or rearranging the original molecular structures. This is done in a series of refining processes. An integrated refinery merely carries on these processes simultaneously, as shown in Figure 3.22-1.

Reduced crude fraction

The highest boiling fraction of crude oil is transferred to a vacuum distillation unit where it is further fractionated. Lubricants, waxes, asphalts, and heavy fuel oils are the end products. The principal hazard is skin contact, with resulting dermatoses. Asphalt, in particular, can be a photosensitizer. A high incidence of scrotal cancer was previously related to the wax preparation process, which has been changed from chilling and mechanical filtration to solvent extraction with toluene. Hazards also include hydrocarbon vapors, heat, and noise.

Heavy gas oil fraction

The heavy oil fraction is transferred to a catalytic cracker where larger molecules are converted to smaller molecules. This is one of the key processes at the refinery and permits doubling of gasoline yields.

The process is operated at about 500°C (930°F) in the presence of a catalyst. There are many commercial catalysts of proprietary compositions, such as alumina-silicates. The catalysts used in this operation are usually nonhazardous. Exposures at the catalytic cracker include hydrocarbons, carbon monoxide from catalyst regeneration, noise, and heat. Heavy oil bottoms containing polynuclear aromatic hydrocarbons are potent skin irritants and potential carcinogens.

The cracked product is next "sweetened" by converting mercaptans to disulfides by treatment with lead oxide, saturated caustic, cuprous chloride, or sodium hypochlorite. The hazards depend on the process being used.

Middle distillate

The middle distillates and raw kerosene are treated in the hydrotreating plant. The organic sulfur compounds are converted to H_2S and removed by burning to SO_2. In addition, hydrocarbons and diolefins are converted to more stable compounds.

The operation is carried out using high pressure hydrogen in the presence of cobalt and molybdenum catalysts and may lead to the formation of toxic metal carbonyl compounds.

Hydrocracking is cracking that takes place when hydrogenation is

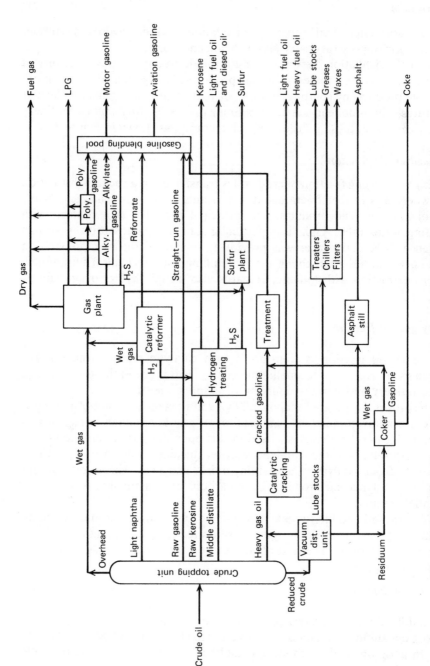

Figure 3.22-1 Flow diagram for refinery. (From Reference 3, Courtesy of American Petroleum Institute)

214

completed at high temperature utilizing a range of catalysts including pelletized tungsten sulfide, tungsten sulfide on granular supports, iron on a special clay support, and nickel or palladium on silica-alumina supports.

Heavy naphtha.

The catalytic reformer produces high octane gasoline, blending components through a rearrangement of molecular structures to highly branched compounds. Thus the quality, not the quantity, of the product is improved during this operation. The procedure is carried out at 480°C (900°F) under high pressure in the presence of either a catalyst such as molybdenum trioxide or aluminum oxide or a bimetal catalyst. Potential exposures during reformer operation include hydrocarbons, hydrogen sulfide, carbon monoxide and noise.

The hydrogen by-product from this process is used in hydrotreating plants. The "wet gas" that is given off is processed further. "Wet gas" is a vapor containing high proportions of hydrocarbons that are recoverable as liquids.

Straight run naphtha

The lowest boiling fraction is first processed in the gas plant, where liquid hydrocarbons found in the wet gas are separated from fuel gases such as propane and butane. Some of the hydrocarbons are run straight through to the gasoline blending plant, but others must be carried through the alkylation process.

In alkylation, isobutane and butylenes or other light olefins are combined in the presence of a hydrogen fluoride or sulfuric acid catalyst. Exposure to the catalysts presents the main hazard of this process.

The final processing involves blending of the various materials to obtain a specific product. There may be exposure to tetraethyl lead where that material is still used and to hydrocarbons from leaks in storage tanks and pump and valve leakage.

Table 3.22-1 is a resumé of the major air contaminants encountered in refinery operations.

Table 3.22-1   *Refinery Air Contaminants*

Principal Air Contaminants	Sources
Hydrocarbon vapors	Transfer and loading operations
	Storage tanks
	Flares
	Cracking unit regeneration
	Boilers
	Pumps, valves
	Cooling towers
	Treating operations

Table 3.22-1 *Refinery Air Contaminants—Continued*

Principal Air Contaminants	Sources
Sulfur dioxide	Boilers
	Cracking unit regeneration
	Flares
	Treating operations
Carbon monoxide	Cracking unit regeneration
	Flares
	Boilers
Nitrogen dioxide	Flares
	Boilers
Hydrogen sulfide	Sour crudes
	Liquid wastes
	Pumps
	Hydrocracker
	Hydrogenation
Particulates	Cracking unit regeneration

3.22.4 Field Studies

Data on comprehensive industrial hygiene studies of refineries in the United States are not available; however, recently, NIOSH sponsored a three-refinery study of worker exposure to carcinogens in fluid catalytic crackers, delayed cokers, and asphalt processing (4). The study included fixed location sampling of 29 polyaromatic hydrocarbons (PAH), 9 aromatic amines, 7 nitrosamines, 6 trace metals, and nickel carbonyl. No detectable quantities of nitrosamines, trace metals, or nickel carbonyl were found. Trace quantities of aromatic amines were found at the fluid cracker. Four to 15 PAH compounds were found in most of the samples. The PAH air concentrations noted in the fluid catalytic crackers ranged from 0–467 $\mu g/m^3$; the delayed coker ranged from 2.9–28.1 $\mu g/m^3$ and the asphalt processing units 0.05–4.9 $\mu g/m^3$.

REFERENCES

1 *International Petroleum Encyclopedia*, Petroleum Publishing Co., Tulsa, 1967.

2 American Petroleum Institute, *Safe Practices in Drilling Operations*, API Recommended Practice, 2010, 3rd ed., API, New York, 1967.

3 American Petroleum Institute, *API Manual on Refinery Hazards*, API, New York, 1978.

4 S. K. Futagaki and D. W. Rumsey, "Industrial Hygiene Characterization of Petroleum Refineries—A NIOSH Study." Paper presented at the American Industrial Hygiene Conference, Houston, May, 1980.

3.23 PLASTICS

There are more than a hundred polymers in production and many hundreds of copolymer systems; however, the polymers listed in Table 3.23-1 represent the bulk of the commonly used materials.

Extrusion molder
- Problems
- Combustion byproducts
- Cleaning & Resins(?)
Nozzle *ETC* *Feed Hopper* *Hydraulic pressure*

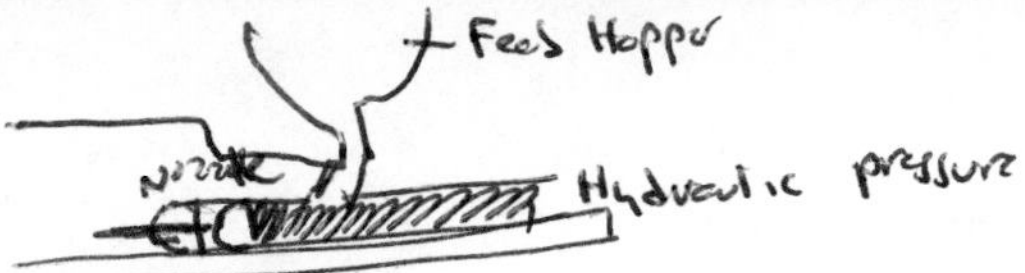

Table 3.23.1 *Chemicals Used in Plastics Production*

Polymer/Resin	Principal Chemicals Used in Production of	
	Monomer	Polymer or Copolymer
Acrylic resins	Hydrocyanic acid Acetone Methyl or isopropyl alcohol	Various catalysts
Alkyd resins	Glycerol Phthalic anhydride Maleic anhydride Linseed, tung, castor oil Litharge Sodium hydroxide Various solvents	Phenol or urea formaldehyde Aliphatic hydrocarbons Styrene Phenols Formaldehyde
Epoxy resins	Allyl chloride Epichorohydrin Diglycidyl ether	Aliphatic amines Organic acids and anhydrides Hydroxy compounds Polyamides
Phenolic resins (or substituted phenol or aldehyde)	Ammonia or lime Hexamethylenetetramine Hydrazine or various amines Phenol Formaldehyde Methyl alcohol	Fillers such as asbestos, various silicate minerals, and cellulose
Amino resins	Ammonia Urea Formaldehyde Methyl alcohol	Same as for phenolic resins
Polyurethane	Ethylene oxide Ethylene glycols Propylene oxide	Toluene diisocyanate (TDI) Hexamethylene diisocyanate (HDI) Diphenylmethane diisocyanate (MDI) Polymethylene polyphenylisocyanate (PAPI) 4,4-methylene-bis-2-chloraniline (MOCA) Organic tin compounds Hydrocarbon blowing agents Polyglycols
Polyamides		Hexamethylene diamine Adipic acid Caprolactum Diphenyl ether-diphenyl mixture
Polyesters	Propylene glycol Maleic anhydride	Organic peroxides Reinforcement such as fiber glass

217

Table 3.23.1   *Chemicals Used in Plastics Production—Continued*

| Polymer/Resin | Principal Chemicals Used in Production of | |
	Monomer	Polymer or Copolymer
	Styrene	Dimethylaniline
	Hydroquinone or *t*-butyl catechol	
Polyolefins	Ethylene	Polyethylene
	Propylene	Polypropylene
	Isobutylene	Polyisobutylene
		Catalysts
		Heavy metals
		Boron trifluoride
		Ammonium chloride
		Aluminum alkyls
Polyvinyl chloride	Hydrogen chloride	Vinyl chloride
	Acetylene or ethylene dichloride	Benzoyl peroxide
		Dioctyl phthalate
	Chlorine	
	Vinyl chloride	
Polystyrene	Benzene	Styrene
	Styrene	

Since plastic production often uses petroleum intermediates as raw materials, as Table 3.23-1 indicates, the production facilities may be located at or near a refinery. The production operations are similar to refinery activities in that the processes are normally conducted in a closed system. In many cases the process is continuous, although batch operations are still common. The large automated plants require relatively few operating personnel.

The plant personnel must tour the plant to check on pump operation, make sight glass readings, and take intermediate and final product samples. The major exposures occurring during such inspections are usually due to fugitive leaks from piping, valves, drains, and pump seals. Maintenance personnel may encounter acute exposure situations.

Many of the polymer production operations are quite straightfoward, neither employing hazardous materials nor resulting in the formation of such materials. The polyolefin plastics are in this "safe" category. In other resin production systems, such as phenol-formaldehyde, the raw materials have known toxicity; however, the facilities usually are totally enclosed and do not present a source of hazardous exposure. In the case of polyvinyl chloride (PVC) the principal exposure to the carcinogenic monomer occurs in the polymer production facility.

A potential problem may be caused by residual monomer in the polymer, as was true of vinyl chloride. Until the 1970s this exposure was significant, but stripping techniques have been introduced to minimize the presence of the unreacted monomer in PVC. The fabricator should be informed of the concentration of the residual monomer in cases of monomer having known toxicity to allow fabrication plants handling the polymer to provide adequate controls.

A principal problem with certain resins such as the epoxies is skin and respiratory sensitization. In the epoxy system the vapor pressure of the resin is very low; thus the resin itself usually does not present an airborne hazard. The curing agent, however, frequently causes dermatitis and sensitization. In the case of polyurethane systems, the isocyanates used as catalysts produce an asthma-like condition.

The production of many low toxicity resins involves the addition of various plasticizers, antioxidants, and stabilizers that are physiologically active and may require control (1). The thermal stability of the parent polymer and the additives are important in considering the application of the product (2). Fire retardant components are among the most recent additives whose toxicity must be considered.

As in the case of refinery processing, a review of the health hazards of plastics production must be based on a flow diagram of the process and data on the raw materials. The initial, intermediate, and final product releases to the atmosphere must be taken into account. Maintenance operations, especially on pumps, reaction vessels, and valves, may offer acute exposures, and specific instructions to protect personnel must be proposed.

REFERENCES

1 R. E. Malten and R. L. Zielhuis, *Industrial Toxicology and Dermatology in the Production and Processing of Plastics,* Elsevier, Amsterdam, 1964.
2 R. A. DeGresero, *Ann. Occup. Hyg.,* **17,** 123 (1974).

3.24 POTTERY

The flow diagram for pottery manufacture shown in Figure 3.24-1 is specifically for the manufacture of table china; however, it includes all the steps encountered in any ceramic manufacturing operation (1). The conventional raw materials include a variety of clays that may contain free silica or quartz, flint that is 100% quartz, and feldspar. These materials are mixed with water in an operation called "blungering" to form "slip." The slip is screened to remove foreign objects and it may be passed over magnets to remove iron particles. The parts are formed by throwing, casting, spreading in a mold, dry pressing, or extrusion. After drying, the rough part is trimmed or "fettled" using knives, sandpaper, or rags. The underside of the ware is sprayed with a mineral oxide coating, dried, and fired in a kiln (box or tunnel furnace). The part is cleaned mechanically or by sandblasting, a final glaze made from frit (crushed glass), clay, and metal compounds is sprayed on, and the part is again fired in a kiln.

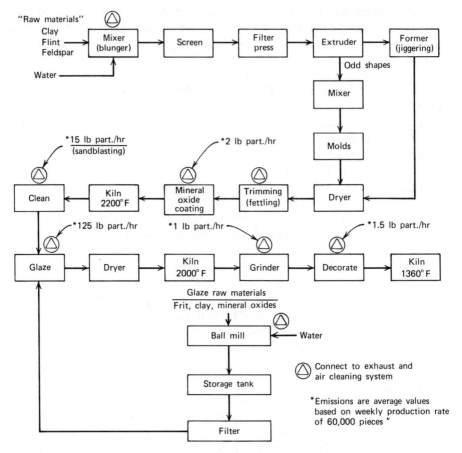

Figure 3.24-1 Flow diagram for pottery operations. (From Reference 1, Courtesy of American Conference of Governmental Industrial Hygienists)

The piece is finished by grinding the base and placing on a decoration decal which is fired in place in a kiln.

The principal hazard in this industry is pneumoconiosis due to the exposure to silica dust. The major exposures to dust occur during the crushing, screening, and preparation of the clay and sand materials, the secondary shaping of the part, and the spraying of the slip and glazes. A major portion of the airborne dust is due to resuspension of dust from equipment, floors, and clothing. The control measures for silica exposure include reducing the percentage of quartz in the raw materials, substituting nonsiliceous material, maintaining all materials in a wet state, provision of local exhaust ventilation, good housekeeping, and personal protective clothing.

The preparation and application of glazes may present a potential hazard due to the use of lead and other heavy metal-based glazes. In the United Kingdom, the lead hazard has been reduced by specifying that the glazes may not contain more than 5% soluble lead. This material specification was backed up by an aggressive program of good housekeeping, clothing changes, locker and shower facilities, prohibition of smoking and eating at the workplace, and local exhaust ventilation at the application locations (2, 3).

REFERENCES

1 Committee on Air Pollution, American Conference of Governmental Industrial Hygienists, "Process Flow Diagram and Air Pollution Emission Estimates," ACGIH, Cincinnati, OH, 1973.

2 Joint Standing Committee for the Pottery Industry, "Dust Control in Potteries," First Report, HM Stationery Office, London, 1963.

3 "Industrial Health, A Survey of the Pottery Industry in Stoke-on-Trent," HM Stationery Office, London, 1959.

3.25 PULP AND PAPER

Pulp is produced by both mechanical and chemical processes. The chemical methods produce more than 80% of the pulp used today. In the kraft, or sulfate, process described in Table 3.25-1 and Figure 3.25-1 (1), chipped wood is digested with steam in tanks using a solution of sodium sulfide and sodium hydroxide (white liquor). Gases are vented periodically from the digester to relieve the pressure buildup. When the digestion is complete, the load is dumped to the blow tank and the gases vent from the pulp and digestion liquid. The spent cooking liquid (black liquor) is drained off, and the pulp is washed, screened, and bleached. The chemicals are recovered from the spent liquor by concentrating it in multiple-effect evaporators. Salt cake is added, and the mixture is sprayed into the recovery furnace; here water is removed, the remaining liquor is burned, and the chemicals are recovered. The chemicals are dissolved in water in the smelt tank, and quicklime is added to convert the sodium carbonate to sodium hydroxide. The calcium carbonate thus formed is converted to calcium oxide in the lime kiln. This product is slaked with water to produce calcium hydroxide for the causticizer (2).

Table 3.25-1 *Kraft (Sulfate) Pulp Process*

Process Step	Formula
White liquor	$NaS_2 + NaOH$
Black liquor	$NaS_2 + NaOH$ + dissolved lignin (15% solids) + Na_2SO_4
Products of combustion of recovery furnace	$NaS_2 + NaCO_3$
Green liquor smelt tank	$NaS_2 + NaCO_3$ in H_2O
Causticizer	$NaS_2 + Na_2CO_3 + Ca(OH)_2$ (quicklime) $\rightarrow NaS_2 + NaOH + CaCO_3 \downarrow$
Lime kiln	$CaCO_3 \xrightarrow{\Delta} CaO + CO_2$
Slaked	$CaO + H_2O \rightarrow Ca(OH)_2$ (for causticizer)

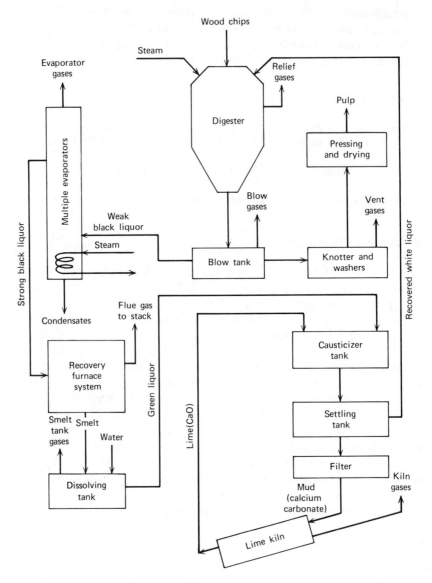

Figure 3.25-1 Flow diagram of pulp and paper plant. (From Reference 1)

The principal exposure to operators occurs when the bottom of the digester is opened and the contents are dumped. The released gases include hydrogen sulfide, methyl mercaptan, dimethyl sulfide, dimethyl disulfide, and sulfur dioxide. The effluent from the recovery furnaces includes organic mercaptans and sulfides, hydrogen sulfide, and sulfur dioxide. (3)

The most significant effort at air pollution control has been the oxidation of the black liquor before the multiple-effect evaporation. In this process, the sulfur compounds are oxidized to produce less volatile materials. Incineration of sulfur off-gases has also been tried by collecting the gases in a gas holder and burning them in the furnace. Other air pollution control techniques are in use that also reduce exposure of workers.

The soda pulp process is similar to the sulfate technique except that sodium carbonate is used for chemical makeup in the furnace. The digestion is carried out with a sodium hydroxide cooking liquor.

In the sulfite process the digester liquor is an aqueous solution of sulfurous acid mixed with lime or other base to form bisulfites. The sulfur dioxide is obtained either as a compressed gas or from the burning of sulfur or the roasting of pyrite ores. The relief gas in this process contains high concentrations of sulfur dioxide, which must be recovered for economical operation. This is accomplished by separators and coolers.

Bleaching of the pulp is usually accomplished with chlorine, followed by extraction with sodium hydroxide, then calcium or sodium hypochlorite, and finally a chlorine dioxide treatment. Chloride hydrate may form when gaseous chlorine enters the vat and is carried to the surface, where it releases chlorine into the atmosphere. As a rule, however, the exposure to chlorine is not difficult to control by local exhaust ventilation.

Paper is coated by coating machines of various types, and the materials used include clay, mica, talc, casein, soda ash, dyes, plastics, gums, varnishes, linseed oil, organic solvents, and plastics. The principal exposures arising from these operations involve acrolein and other aldehydes resulting from the atmospheric oxidation of linseed oil and solvent vapors from the coating and subsequent drying of the paper. When the coating and drying are done in air-conditioned rooms, the environmental control problems become difficult.

Lime exposure may be excessive in both the sulfate and soda processes during handling of the lime. The digestion pit and the first washing cycle should be exhausted to eliminate sulfur dioxide, hydrogen sulfide, and mercaptans. A serious sulfur dioxide exposure may occur throughout the entire sulfate process. This gas can be controlled by ventilation at the sulfur burners, thereby achieving negative pressure on the acid towers, ventilation on the digesters, and remote operation of blowdown valves.

REFERENCES

1 "Air Pollution and the Kraft Pulping Industry, An Annotated Bibliography," U.S. Department of Health Education and Welfare, Public Health Service Publication No. 999-AP-4, Cincinnati OH, 1963.

2 R. G. MacDonald and J. N. Franklin, Eds., *Pulp and Paper Manufacture*, 2nd ed., Vol. 3, McGraw-Hill, New York, 1969.

3 S. S. Gautam, A. V. Venkatanarayanan, and B. Parthasarathy, *Ann. Occup. Hyg.*, **22**, 371 (1979)

3.26   RAYON

In the rayon-making process, pads or sheets of cellulose prepared from wood pulp are steeped in sodium hydroxide to form a "soda" cellulose. The sheet material is shredded, aged, and mixed with carbon disulfide in an xanthating churn. The cellulose xanthate, in solution in alkali, is a brown syrupy liquid known as viscose. After filtration, aging, and deaeration, the viscose is forced through small holes in a nozzle or "spinerette" submerged in a sulfuric acid bath. The stream of viscose emitted from this spinning operation contacts the bath, and the cellulose is regenerated to form a continuous fiber. Tension on the fiber is established using two rollers called "godets," which operate at slightly different speeds. This process orients the yarn fibers in position parallel to the yarn axis. The yarn may be chopped into short elements at this point for additional processing.

The principal hazard in this industry is the exposure to carbon disulfide in the xanthation, spinning, "godet," and cutter house operations (1). The xanthation process usually can be controlled, since it is an enclosed operation. The open spinning baths release carbon disulfide and hydrogen sulfide, and operations should be monitored and ventilation control applied. The stretching and processing of the fiber at the "godets" is a principal exposure point because carbon disulfide is released from the fiber. Moreover, if the continuous fiber breaks, the fiber or "tow" must be pulled manually, resulting in a serious carbon disulfide exposure. This operation must be controlled by local exhaust ventilation utilizing an enclosing hood with a face velocity of 1.0 m/sec (200 fpm).

Extensive air sampling data are available on the concentration of carbon disulfide and hydrogen sulfide in viscose rayon plants. In a Finnish plant, the concentrations of carbon disulfide were approximately 10 times those of hydrogen sulfide. The combined concentrations, which were greater than 40 ppm before 1950, had dropped to below 5 ppm by 1972 (2). A NIOSH Health Hazard Evaluation demonstrated that concentrations of carbon disulfide routinely exceeded 20 ppm as an 8 hr TWA concentration in a rayon fiber plant (3). In another survey at viscose rayon plants in the United States, carbon disulfide concentrations in the churn and spinning rooms were 10 to 15 ppm (4).

The complete range of conventional controls must be employed in the rayon industry to minimize exposure to carbon disulfide. Work practices must be carefully defined; housekeeping and handling procedures must be encouraged to reduce the deposits of waste viscose, bath solution, and tow on the floor; it must be recognized that effective ventilation controls on spinning and cutter house operations are essential. Medical control by periodic examination of workers and biological monitoring is necessary. If work practices and engineering and administrative controls are not adequate, a comprehensive respirator program must be implemented.

REFERENCES

1 "Criteria for a Recommended Standard—Occupational Exposure to Carbon Disulfide," U.S. Department of Health, Education and Welfare, Publication No. (NIOSH) 77-156, Cincinnati, OH, 1977.

2 S. Hernberg, T. Partanen, C. H. Nordman, and P. Sumari, *Br. J. Ind. Med.*, **27**, 313 (1970).

3 Health Hazard Evaluation Determination Report No. 72-21-91, U.S. Department of Health, Education and Welfare, NIOSH, Cincinnati, OH, 1973.

4 Plant observation reports and evaluations for carbon disulfide, Stanford Research Institute, NIOSH Contract No. CDC-99-74-31, Stanford, CA, 1977.

3.27 RENDERING PLANTS

Rendering plants process meat and bone scrap to produce tallow and protein meal. The reduction is usually accomplished in a batch operation as shown in Figure 3.27-1 (1). The cookers are steam jacketed horizontal cylinders that will handle 2700 to 5400 kg (6000 to 12,000 lb) of material.

Material received at the plant is dumped to a recovery pit and, when required, it is processed through a breaker to a charging bin over the cookers. The cooked mixture is dumped, and the tallow drains off by gravity. Solids are pressed and then ground for meal product use. The tallow is processed through either a filter press or centrifuge.

Malodors from rendering plants are emitted principally from the cookers, driers, air blowing of tallow, and percolating pans. The organic material with the offensive odors are aldehydes, fatty acids, amines, mercaptans, and sulfides. The odor threshold levels for many of the rendering plant emissions

Table 3.27-1 *Odor Threshold Levels for Selected Compounds*

Compound	Chemical Formula	Odor Threshold (ppb)
Dimethyl amine	Ch_3NHCH_3	4.7
Methyl amine	CH_3NH_2	21.0
Trimethyl amine	$(CH_3)_3N$	0.21
Ammonia	NH_3	46,800
Ethyl mercaptan	C_2H_5SH	1.0
Hydrogen sulfide	H_2S	4.7
Methyl mercaptan	CH_3SH	2.1
Dimethyl sulfide	CH_3SCH_3	2.5
Dimethyl disulfide	CH_3SSCH_3	7.6
Skatole	C_9H_8NH	220
Acrolein	$CH_2=CHCHO$	210
Butyric acid	C_3H_7COOH	1.0

Source: Reference 2

are shown in Table 3.27-1 (2). Although the air concentrations for these contaminants prompt worker and neighborhood odor complaints, normally they will be far below any occupational health standards. Control must be achieved by collecting all emissions and providing air cleaning by scrubber or incinerator.

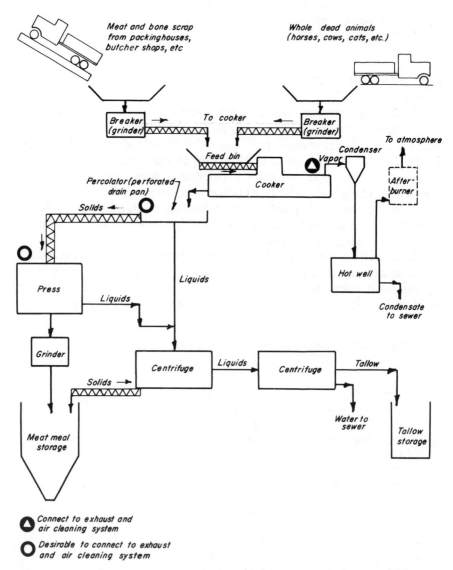

Figure 3.27-1 Flow diagram for inedible rendering plant—standard process. (From Reference 1, Courtesy of American Conference of Governmental Industrial Hygienists)

REFERENCES

1 Committee on Air Pollution, American Conference of Governmental Industrial Hygienists, "Process Flow Diagram and Air Pollution Emission Estimates," ACGIH, Cincinnati, OH, 1973.
2 T. Osag and G. B. Crane, "Control of Odors from Inedibles-Rendering Plants," U.S. Environmental Protection Agency Publication No. EPA-450/1-74-006, Research Triangle Park, NC, July 1974.

3.28 RUBBER

The first man-made rigid polymer, natural rubber, has been joined by at least two dozen synthetic rubber polymers used in a variety of industrial applications (1). The commonly used rubber polymers and copolymers are given in Table 3.28-1. Many applications require a blending of natural and synthetic rubbers to take advantage of the unique characteristics of each. This is the case in tire manufacture, which represents the lagest single rubber application. Natural rubber is converted at its source to dry rubber or to a latex concentrate. In either case, a preservative such as ammonia, formaldehyde, or sodium sulfite, is added. The material is shipped in bales or in barrels. The synthetic polymers are received in solid bale form.

Table 3.28-1 *Types of Rubbers in Order of Appearance*

Natural rubber
Polysulfide polymers
Polychloroprene
Nitrile rubber
Styrene-butadiene rubber
Butyl rubber
Polybutadiene
Silicone rubber
Acrylic rubber
Chlorosulfonated polyethylene
Polyurethanes
cis-Polybutadiene
cis-Polyisoprene
Fluorine-containing elastomers
Epichlorohydrin elastomers
Ethylene-propylenediene elastomers

Source: Reference 1.

Various rubber processing chemicals are presently in use to permit fabrication into finished product and to ensure specific properties for the product. Usually organic materials, these chemicals are added in relatively small quantities to the rubber stock formula. The chemicals and production facilities employed in processing these rubber materials are similar and can be discussed in a general fashion.

3.28.1 Materials

Vulcanizing Agents

Still the most important vulcanizing agent, sulfur is used either as elemental sulfur or in one of many organic forms. The common vulcanizing materials appear in Table 3.28-2.

Table 3.28-2 *Rubber Vulcanizing Agents*

Tetramethylthiuram disulfide
Tetrathiuram disulfide
Dipentamethylene thiuram tetrasulfide
4,4-Dithiodimorpholine
Selenium diethyldithiocarbamate
Aliphatic polysulfide polymer
Alkylphenol disulfides

Source: Reference 1.

Accelerators

Since the early 1900's, chemicals have been added to rubber systems to hasten vulcanization. Initially, inorganic lead compounds were tried, then aniline and finally a series of various organic compounds were used. The principal accelerators in current use are listed in Table 3.28-3. At present, the thiazole accelerators such as benzothiazole sulfenamides are most common.

Table 3.28-3 *Commercial Accelerators*

Aldehyde-amine reaction products
Arylguanidines
Dithiocarbonates
Thiuram sulfides
Thiazoles
Sulfenamides
Xanthates
Thioureas

Source: Reference 1.

Activators

Zinc oxide and fatty acid commonly are used in conjunction with accelerators to achieve a given property. Other activators such as litharge, magnesium oxide, amines, and amine soaps are also employed.   Lead

Antioxidants

Stabilizers are designed to protect the polymer during extended storage before its end use in manufacturing, whereas antioxidants are designed to protect the finished product. The important antioxidants are arylamines and phenols, as given in Table 3.28-4.

Table 3.28-4 *Commercial Antioxidants*

Arylamines
   Aldehyde-amines
   Aldehyde-imines
   Ketone-amines
   p-Phenylenediamines
   Diarylamines
   Alkylated diarylamines
   Ketone-diarylamines
Phenols
   Substituted phenols
   Alkylated bisphenols
   Substituted hydroquinones
   Thiobisphenols

Source: Reference 1.

Antiozonants

The principal antiozonants are symmetrical *p*-phenylenediamine, unsymmetrical *p*-phenylenediamines, dihydroquinolines, and dithiocarbonate metal salts. Paraffin and microcrystalline waxes also are used in rubber components.

Plasticizers

The viscosity and therefore the workability of the polymers can be improved by adding organic lubricants or physical softeners to the rubber. These additives may include coal tar, petroleum, ester plasticizers, liquid rubbers, fats and oils, and synthetic resins.

Pigments

Pigments in rubber occasionally have as their principal contribution the addition of color, but usually they are applied as reinforcing pigments, fillers, or extenders. The common pigments are carbon black, zinc oxide, clay, and silicates.

3.28.2 Processes

The processing techniques are similar throughout the rubber industry. It is common to have a defined processing area where weighing and processing are conducted. A normal tread stock recipe (Table 3.28-5) gives an idea of the quantities of materials involved. In some cases the bulk natural or synthetic polymer must be worked in a breakdown or mastication mill to make the stock more flexible and easy to work. The various rubber chemicals specified in the formula are then weighed out for a specific batch size. This operation involves opening bagged and barreled material and placing it in hoppers for weighing. Since most of the compounds are solid and granular, this usually is a dusty operation, and weigh stations should be exhausted. Where possible, small quantities of chemicals should be prepackaged in plastic bags that can be placed directly in the batch, thereby eliminating a dust exposure in emptying bag contents. The manufacturing steps in tire production shown in Figure 3.28-1 are representative of most rubber products as are the occupational titles shown in Table 3.28-6 (2, 3). The individual components are either mixed in an open two-roll mill or an internal mixer such as a Banbury. In either case, dust is released during charging to the mixer. Standard local exhaust designs are available for both types of equipment, and such control should be installed (4). Of the two processes, better control can be established on the Banbury, since a more effective enclosure can be fabricated.

Compounding

Mixing

Table 3.28-5 *Natural Rubber Tire Tread Compounds*

Material	Parts
Natural rubber	100
Sulfur	3
Accelerator	1
Zinc oxide	5
Stearic acid	1
Antioxidant	1
Softener	5
Pigment	50
Total	166

Source. Reference 1.

The rubber batch is processed from the Banbury to a drop mill or from the mixing mill to a sheeting mill. During this transfer the hot stock may release volatile fractions of oils, high vapor pressure organic components, or possibly degradation products of the worked material. If biocidal additives are in the formula, they may be released. At the drop mill the stock is blended, sheeted, cooled, and cut for racking.

At this point the stock is usually dusted or dipped in a talc slurry or other material to reduce sticking. If talc or soapstone is used, it should be analyzed to ensure that it does not contain asbestos and that it has negligible free silica. The wet slurry process is obviously the most effective way to add the talc, since the level of dustiness is much lower than that for dry dusting.

Table 3.28-6 *Description of Occupational Title Groups in Tire and Tube Manufacturing*

Occupational Title Group	Description of Process
Compounding	Batch lots of rubber stock ingredients are weighed and prepared for subsequent mixing in Banburys; solvents and cements are prepared for process use.
Banbury mixing	Raw ingredients (rubber, fillers, extender oils, accelerators, antioxidants) are mixed together in a Banbury mixer. This internal mixer breaks down rubber for thorough and uniform dispersion of the other ingredients.
Milling	The batches from the Banbury are mixed further on a mill and cooled; the sheets or slabs are coated with talc so they are not tacky. The stock may return to the Banbury for additional ingredients or go on to breakdown or feed mills before extrusion or calendering.
Extrusion	The softened rubber is forced through a die forming a long, continuous strip, in the shape of tread or tube stock. This strip is cut in appropriate lengths, and the cut ends are cemented so they are tacky.
Calendering	The softened rubber from the feed mill is applied to fabric forming continuous sheets of plystock in the calendar (a mill with three or more vertical rolls and much greater accuracy and control of thickness).
Plystock preparation	The plystock from the calendar is cut and spliced to the correct size for tire building, and so the strands in the fabric have the proper orientation.
Bead building	Parallel steel wire is insulated with rubber vulcanizable into a semihard condition and covered with a special rubberized fabric. The beads maintain the shape of the tire and hold it on the wheel rim in use.
Tire building	The tire is built from several sheets of calendered plystock, treads and beads.
Curing preparation	The assembled green or uncured tire is inspected, repaired, and coated with agents to keep it from sticking to the mold in vulcanization.
Tube splicing	Assembly of tube stock, that is, tube building.
Curing	The green tire or tube is placed in a mold and vulcanized under heat and pressure.
Final inspection and repair	The cured tire is trimmed, inspected, and labeled; repairable tires or tubes which do not pass initial inspection are repaired.

Source. Reference 2.

[Handwritten margin annotations: "High Dust", "Preparing the raw materials", "Heat stress", "Accidental pull-in"]

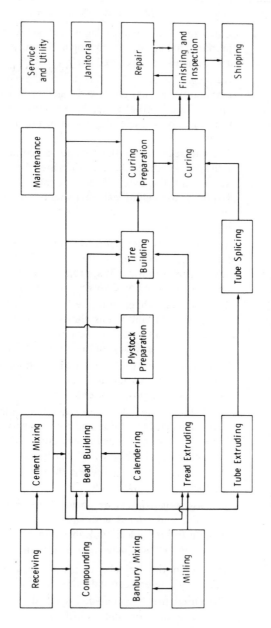

Figure 3.28-1 Flow diagram of tire and tube manufacture. (From Reference 2, Courtesy of American Industrial Hygiene Association)

This batch can include the vulcanizing agent or, if the rubber is to be stored, a master batch can be produced that includes all components except a part or all of the curing agent. The basic rubber stock is then ready for a variety of processes. Specific shapes, such as tire treads, may be formed by extrusion. Dimensional rubber sheeting may be produced using a multiroll calender, which smears the rubber into a fabric to make reinforced rubber sheets such as tire ply. If materials are "layed up" for sheet fabrication, as in the case of tarpaulins or clothing, exposure to rubber cement solvents may occur. The type of solvent depends on the application. Ventilated work benches are usually required for these operations (5). At one time the use of benzene was common and constituted a significant hazard, but now white gasoline-type solvents are commonly substituted for aromatic hydrocarbons. In recent years the aromatic content of white gasoline has been increased, and this factor should be monitored by air sampling. Plants should specify that white gasoline contain less than 0.1% benzene.

The assembled products usually require curing with heat. These procedures vary and may present specific hazards. Tires are cured in molds, with the aluminum molds heating the outside and water- or steam-filled bladders heating the inside of the tire. Large conveyor belts are cured in flat, steam-heated presses. Water hose is cured by extruding a lead sheath over the hose and directing steam inside to cure the part. Continuous bath and drum curing are also conducted. Certain problems are common to all curing operations: release of curing fume, whose composition and concentration depend on the rubber stock and the curing temperature, heat stress due to the steam release and convective load, and noise exposure from air and steam release.

3.28.3   Field Studies

A comprehensive survey of airborne contaminants in U.S. tire plants has been completed by the Occupational Health Studies Group at the School of Public Health of North Carolina under the sponsorship of the United Rubber Workers Union and six major rubber products manufacturing companies. Particulate concentrations in air were evaluated in 14 tire and tube manufacturing plants and these data are summarized in Table 3.28-7; solvent concentrations shown in Table 3.28-8 (2, 3) represent conditions in 10 tire manufacturing plants. The authors state that these values are the aggregate of all survey data and do not represent time-weighted averages.

Hexane is highest
↓

Table 3.28-7 *Air Concentrations of Solvents by Occupational Title Group (OTG)*

OTG	Concentration (ppm)[a]				
	Pentane	Hexane	Heptane	Benzene	Toluene
Cement mixing					
Personnel	3.9(7)	15.7(8)	2.8(8)	0.5(8)	3.1(8)
Area	0.9(8)	9.5(9)	4.7(7)	1.0(8)	2.9(8)
Extrusion					
Personnel	1.4(9)	5.9(9)	13.6(9)	0.8(9)	9.5(9)
Area	5.4(4)	12.2(4)	10.2(4)	1.2(4)	14.0(4)
Tire building					
Personnel	1.8(10)	11.2(10)	5.7(9)	1.4(10)	1.5(10)
Area	5.8(2)	19.3(3)	14.6(3)	1.9(2)	8.0(2)
Curing					
Personnel	0.8(8)	7.5(9)	4.9(7)	0.8(9)	1.2(9)
Area	1.8(2)	25.9(3)	7.5(2)	0.8(2)	0.6(3)
Inspection and repair					
Personnel	1.5(3)	6.0(3)	3.1(3)	1.1(3)	0.8(3)
Area	1.3(3)	6.6(3)	4.2(3)	0.6(3)	1.9(3)

Source. Reference 2.

[a]Column entries are the middle value, or mean of two middle values, of plant arithmetic mean concentrations, and the number of plant means (in parentheses).

Table 3.28-8 *Air Concentration of Respirable Particulates by Personal Sampling*

OTG	Concentration (mg/m³)[a]		
	Median	(no. of plants)	Range
Compounding	0.69	(9)	0.39–1.57
Banbury	0.64	(10)	0.33–1.47
Milling	0.48	(10)	0.10–0.88
Tread extrusion	0.53	(2)	0.14–0.91
Calendering	0.28	(1)	
Bead building	0.24	(2)	0.23–0.24
Tire curing	0.43	(3)	0.03–1.0
Tire finishing, inspection, and repair	0.22	(11)	0.04–0.58
Tube curing	0.65	(7)	0.29–1.67

Source. Reference 3.

[a]Column entries are the middle value, or mean of two middle values, of plant arithmetic mean concentrations, the number of plant means (in parentheses), and the range of plant mean concentrations.

REFERENCES

1 H. F. Mark, N. G. Gaylord, and N. Bikales, Eds., *Encyclopedia of Polymer Science and Technology*, Vol. 12, *Plastics, Resins, Rubbers, Fibers*, Wiley-Interscience, New York, 1970.

2 T. M. Williams, R. Harris, E. Arp, M. Symans and M. VanErt, *Am. Ind. Hyg. Assoc. J.*, **41**, 204 (1980).

3 M. VanErt, E. Arp, R. Harris, M. Symans, and T. Williams, *Am. Ind. Hyg. Assoc. J.*, **41**, 212 (1980).

4 Committee on Industrial Ventilation, American Conference of Governmental Industrial Hygienists, *Industrial Ventilation: A Manual of Recommended Practice*, 14th ed., ACGIH, Lansing, MI, 1976.

5 British Rubber Manufacturers Association, "Toxicity and Safe Handling of Rubber Chemicals, BMRA Code of Practice," BMRA, London, England, 1978.

3.29 SHIPBUILDING AND REPAIR

The modern shipyard is a complex facility incorporating machining, welding, founding, painting, electroplating, abrasive blasting, and electronic repair operations (1, 2). The control of occupational health problems is complicated because the activities are not confined to the shops but are often carried out on shipboard. The unit operations conducted at the various shipyard service shops are similar to general industry and the hazards and controls for these operations described in Chapter 2 apply to the shipyard environment. Aboard ship, however, several trades may carry out simultaneously new construction or repair operations in a small confined space such as an engine room or boiler room. In these situations, all tradespersons are exposed to the contaminants generated by the various operations.

Before a vessel enters a dry dock for construction work, repairs, or alterations of any kind, all tanks, compartments, or lines that have contained flammable liquids should be cleaned and freed of flammable vapor. The atmosphere in all unventilated areas or compartments should be checked for harmful or flammable gases and for oxygen deficiency by an industrial hygienist or qualified marine chemist. All provisions for safe entry to confined spaces should be implemented. Shipyards must have specific instructions for surface preparation, coating, and welding or "hot work" (3).

Tankers that have carried gasoline or volatile crude oils require periodic checks even after a "gas-free" status has been established. Rust on the bulkheads or decks of compartments may continue to dissipate flammable vapor. The pumping of ballast also may introduce flammable vapor from some inaccessible part of the pipe lines or storage tanks. When work is conducted in tanks used for the transport of gasoline, the lead exposure involved in welding or cutting operations on rust-coated surfaces must be evaluated.

Asbestos is used routinely for insulation of steam lines in high pressure marine power plants. Repair of pipe lines during shipyard availability includes "rip out" of the insulation. This job is extremely dusty and unless proper controls are established airborne fiber concentrations may exceed 100 fibers/ml for short periods. Limited data from a shipboard asbestos work project are shown in Table 3.29-1 (4). Several papers have been written on the control technology to minimize exposure to asbestos. The space should be isolated and all other trades excluded from the area. Personnel doing the "rip out" should be equipped with respiratory protection. Exposure concentrations may require continuous flow air-line respirators or powered air-purifying respirators. The insulation is infused with water and removed in as large a piece as possible. The debris is placed in plastic-lined barrels or drums for disposal. Ventilation control on large jobs should include a portable exhaust fan equipped with a high-efficiency particulate filter. Vacuum cleaners designed for handling asbestos should be used continuously for cleanup. Air hoses and brooms should not be allowed in the work space. Drapes and floor covering are useful in confining the waste and facilitating cleanup. If possible, the insulation should be replaced with nonasbestos insulation. If asbestos insulation must be used, the insulation elements should be fabricated in a shop facility equipped with adequate dust controls. Premixed mud should be used during installation. If possible, nonasbestos cloth should be used for pipe covering.

Table 3.29-1 *Asbestos Dust Exposure at Various Operations in Shop and on Board Ship*

Location	Operation	Number of Air Samples	Number of Samples greater than 5 f/ml	Number of Samples greater than 12 f/ml	Concentration Range Minimum (fibers/ml)	Concentration Range Maximum (fibers/ml)	Mean Concentration (fibers/ml)
Shop	General air	4	0	0	0.1	0.1	0.1
	Bench work	12	0	0	0.1	0.3	0.2
	Cutting on band saw	3	0	0	0.1	3.1	1.1
Ship	General air, no active work	4	0	0	0.0	0.1	0.1
	Installation of insulation	27	5	1	0.1	15.3	3.6
	Tearing out old insulation	25	6	1	0.1	25.0	3.1
	Mixing cement	1	0	0	—	0.4	—

Source. Reference 4.

The most widespread exposures in shipbuilding and repair are those connected with welding and cutting. There are many opportunities for welding in confined spaces on a ship, areas into which a person must crawl and where there is no ventilation except what is supplied mechanically. Under such circumstances the air concentration of nitrogen dioxide produced by a gas torch can reach fatal levels in a matter of minutes. Properly distributed mechanical ventilation is the control method of choice, complemented by supplied-air respirators.

As discussed under welding in Chapter 2, a major hazard results from welding and cutting on surfaces coated with lead-and chromium-based finishes, antifouling hull paints, and galvanized zinc coating. If possible, the offending materials should be removed mechanically by scraping, wire brushing, or blasting before welding or cutting. If that is not possible, local exhaust ventilation and respiratory protection must be used to protect the worker during welding and cutting.

Spray painting of tanks and compartments requires not only control of inhalation exposures to thinners, oils, and pigments, but also prevention of fire and explosion. Control can be accomplished by the application of mechanical air supply and exhaust (5). Where necessary, this can be supplemented by air-line or air-purifying respirators, depending on the nature and the level of air contamination. Painting of the hull poses problems not only of respiratory protection but of visibility, because uncontrolled paint mist quickly covers goggles or face shields. Where natural or artificially-induced air movement cannot be used to advantage to prevent the inhalation of paint mists, spray nozzles mounted on long pipes are effective in minimizing exposure.

A wide range of antifouling paints are applied to ship hulls. During the spray application of these materials one is exposed to the solvent system, which may contain aliphatic or aromatic hydrocarbons, and various ketones. The fungicidal materials used during the past decade include organo-mercury compounds, copper oxide, arsenic, and, more commonly in recent years, organo-tin compounds such as dibutyltin-oxide (6).

Abrasive blasting is conducted with a variety of abrasives. The experience in Louisiana yards reemphasizes the hazard from the use of sand in blasting. In the period from 1960 to 1975, 100 cases of silicosis were diagnosed in a work population blasting Texas Tower structures in Louisiana shipyards. A total of 25 deaths was chronicled (7). The average duration of exposure for the sandblasters was 10 years. Other investigators have cited silicosis from crystalline silica after exposure periods as brief as 18 months. The obvious control is to use other less toxic abrasives. If sand must be used, and that seems unlikely, an effective respiratory protection program based on air-supplied blasting helmets must be implemented.

To evaluate the hazardous nature of ship cargoes and to establish the proper precautions to be exercised in their handling requires a rather broad understanding of industrial toxicology. Oxygen deficiency resulting from fermentation, dry ice refrigeration, or displacement of air by gases other than carbon dioxide has probably caused more fatalities on cargo ships than any atmospheric contamination except explosions of flammable vapors. Skin irritation from cargoes is also common.

REFERENCES

1 B. G. Ferris, Jr., and H. Heimann, *Environ. Res.* **11**, 140 (1976).

2 Part 1502 Safety and Health Regulations For Ship Repairing, Excerpt From Code of Federal Regulations, Title 29, U. S. Department of Labor, Washington, D.C., 1971.

3 "Hot Work: Welding and Cutting on Plant Containing Flammable Materials," U.K. Publication HS(G)5, Health and Safety Executive, London, 1979.

4 W. A. Burgess, "Exposure of Insulation Workers At Boston Naval Shipyard," unpublished data.

5 Bethlehem Steel Corporation, Shipbuilding Dept., "Manual of Recommended Practices For Ventilation of Coating and Surface Preparation Operations," Bethlehem, PA, 1968.

6 O. D. Llewellyn, *Ann. Occup. Hyg.*, **15**, 393 (1972).

7 B. Samimi, H. Weill, and M. Ziskind, *Arch. Environ. Health*, **29**, 61 (1974).

3.30 SMELTING

This discussion will cover only primary smelting of copper and lead. Smelting of other metals present similar occupational health hazards.

3.30.1 Copper

The principal processes in copper refining include mining, concentrating, and smelting (1). Copper ore is available as either an oxide or sulfide ore (Table 3.30-1) and usually is surface mined. Ore beneficiation is conducted near the mining site by grinding the crushed ore in a ball mill to form a slurry to which flotation "assists" are added. The froth containing the copper is skimmed off, dewatered, and shipped to the smelter as a 16–32% concentrate with approximately the composition shown in Table 3.30-2. The workplace exposures in a copper smelter are identified in Table 3.30-3 and the major steps are portrayed in Figure 3.30-1.

Table 3.30-1 *Composition of Copper Ores*

	Oxides		Sulfides
Malachite	$CuCO_3 \cdot Cu(OH)_2$	Chalcocite	Cu_2S
Azurite	$2\ CuCO_3 \cdot Cu(OH)_2$	Chalcopyrite	$Cu_2S \cdot Fe_2S_3$
Cuprite	Cu_2O	Covallite	CuS
Atacamite	$CuCl_2$	Bornite	$FeS \cdot 2\ Cu_2S \cdot CuS$
Brochantite	$CuSO_4 \cdot 3\ Cu(OH)_2$		

Source. Reference 1.

Table 3.30-2 *Composition of Copper Ore*

Element	Percent
Copper	16–32
Arsenic	0.001–6.7
Lead	0.003–1.3
Zinc	0.004–2.2
Cadmium	0.001–0.04
Molybdenum	0.1–0.5

Source. Reference 1.

The initial operation at the smelter is roasting, which dries the ore concentrate and controls the sulfur content. The calcine that is formed is fed to the smelters with recycled precipitates, converter slag, flue dust, limestone, and silica flux. Matte, a mixture of cuprous and ferric sulfide, is formed along with a slag of various metal silicates. The matte, which contains 35% Cu, is charged by ladle to the converter with siliceous flux. Air is blown into the hot

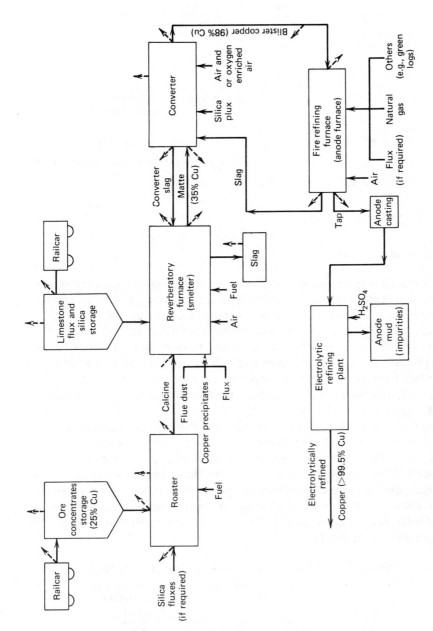

Figure 3.30-1 Process flow diagram for primary copper smelting showing potential industrial process fugitive emission points. (From Environmental Protection Agency)

239

Table 3.30-3 *Copper Smelting Reactions and Exposures*

Operation	Purpose	Equipment	Reactions	Workplace Exposures
Roasting	Dries ore concentrate Controls S content Produces calcine	Roasters multiple hearth fluid bed	$2\,Cu_2S + 3O_2 \rightarrow 2Cu_2O + 2SO_2$ (calcine)	Ore dust SO$_2$ CO Heat
Smelting	Produces Cu–Fe sulfide matte (35% Cu) and siliceous slag Charge is concentrate or calcine, recycled precipitates, converter slag, flux dust, limestone, and silica flux	Furnace reverberatory electric furnace	$Cu_2S + 2CuO \rightarrow 4Cu + SO_2$ $2Cu + FeS \rightarrow Cu_2S + Fe$ (matte)	Lead containing dust (8%) Flux dust CO Matte dust/fume SO$_2$
Converting	Produces blister copper (98.5%) Charge is matte and silica flux	Converter	$2Cu_2S + 3O_2 \rightarrow 2Cu_2O + 2SO_2$ $Cu_2S + 2CuO \rightarrow 4Cu + SO_2$ (blister Cu)	Lead containing dust (1–8%) Flux dust CO SO$_2$ Metal dust/fume
Refining	Produces domestic copper (>99.5% Cu)	Electrolytic refining	Electrolytic bath of $CuSO_4 + H_2SO_4$ $Cu(impure) \rightarrow Cu_{2+} \rightarrow Cu$ (pure) Impurities come off as slimes	H$_2$SO$_4$ mist

metal through tuyeres and metal silicate slag is formed on top of the cuprous
sulfide. The slag is recycled periodically to the smelting operation to recover
the remaining copper. The blister copper formed in the converter is
approximately 98.5% pure. This intermediate is further refined in a gas-fired
furnace and the copper is poured into molds to form slabs of blister copper.
This product is refined electrolytically in a copper sulfate-sulfuric acid
electrolyte bath with the blister copper as the anode and a thin plate of pure
copper as the cathode. The copper from the blister copper slab is deposited as
99.5% copper on the cathode.

Since ore, flux, and by-product granular material are handled in large
quantities in copper smelting, there is a dust exposure to siliceous dust and
metal fumes. In almost all operations, sulfur dioxide is present, as is carbon
monoxide from the combustion processes. Noise and heat stress are the
principal physical hazards, the latter due primarily to radiation from furnaces.
Representative data on sulfur dioxide and metal fume concentrations are
presented in Tables 3.30-4 and 3.30-5 (2). The metal fume exposures depend on
the composition of the feedstock.

Table 3.30-4 *Sulfur Dioxide Concentrations, Area Samples*

Area	Concentration (ppm)
Reverberatory furnace charging deck	13.2
Reverberatory furnace operators deck	4.4
Converter	4.0
Anode casting	1.3

Source. Reference 2.

Table 3.30-5 *Concentrations in Air, Personal Sampling* [a]

			mg/m³			
Area	Pb	Zn	ZnO (calculated)	Cu (dust and fume)	Cd (dust and fume)	Mo (total)
Reverberatory Furnace Charging deck	0.07	0.12	0.15	3.4	0.005	0.003
Reverberatory Furnace Operators deck	0.07	0.07	0.09	1.3	0.006	0.03
Converter	0.03	0.04	0.05	0.11	0.004	No data
Anode casting	0.01	0.01	0.01	0.07	0.001	No data
Current OSHA standard	0.05	—	5	Dust 1.0 Fume 0.1	Dust 0.2 Fume 0.1	Soluble 5 Insoluble 15
NIOSH recommended standard	0.15	—	5	Dust 0.04	Fume 0.04	—

Source.   Reference 2.
[a]Industry-wide, not representative of any one location.

Dust emission is controlled by local exhaust ventilation at material transfer points; sulfur dioxide exposures are controlled by both local exhaust and dilution ventilation. Respirators for particulate and acid gas protection are worn routinely.

3.30.2   Lead

The processing of lead includes all common smelting and refining techniques. The flow sheet for lead smelting is shown in Figure 3.30-2 and the workplace exposures are tabulated in Table 3.30-6 (3).

Table 3.30-6 *Lead Smelting Reactions and Exposures*

Operation	Purpose	Equipment	Reactions	Workplace Exposures (emission concentration)
Sintering	Convert sulfides to oxides and sulfates	Sintering machine 1000°C (1830°F)	$2\ PbS + 3\ O_2 \rightarrow 2\ PbO + 2\ SO_2$	SO_2 (0–6.5% in stream) Lead containing dust (20–65%)
Smelting	Removes impurities, reduces compounds to Pb (bullion) containing 94–98% Pb and slag	Lead blast furnace	$2\ PbO + 2\ C \rightarrow 2\ Pb + 2\ CO$ $PbO + CO \rightarrow Pb + CO_2$	CO (2%) SO_2 (0.01–0.25%) Siliceous dust Pb dust Other metallic oxides
Drossing	Remove Cu, S, As, Sb, Ni from solution	Dross kettles	Various	Impurities in bullion Cu, Sn, Bi, As, Cu CdO, Sb CO SO_2 Pb dust

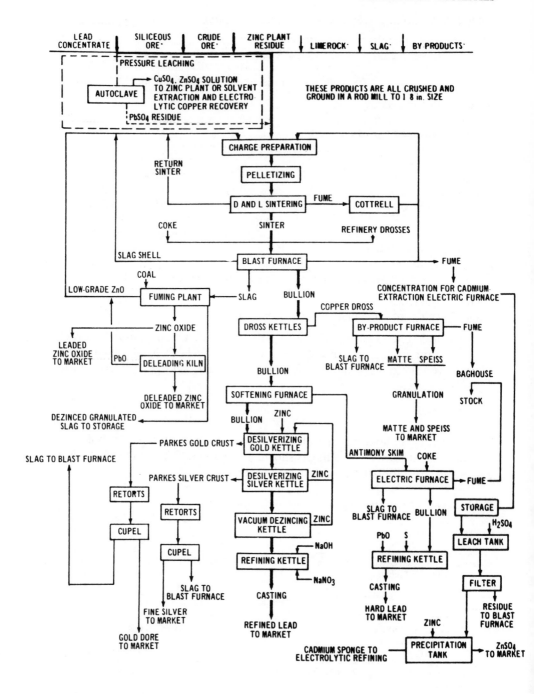

Figure 3.30-2 Typical flow sheet of pyrometallurgical lead smelting. (From Environmental Protection Agency)

The crushed and ground ore, which may contain 3 to 8% lead along with other heavy metals, is concentrated using differential flotation. The concentrate is mixed with limestone, silica sand, and iron ore, and is then pelletized. The pelletized product is sintered and in the process the sulfur content is reduced by oxidation to sulfur dioxide and the lead sulfide is converted to the oxide. The sulfur dioxide gas is recovered by an acid plant. Sinter coke, limestone, silica, litharge, and slag-forming materials are charged to a blast furnace for smelting, and the lead is reduced to metallic lead. The molten lead and slag are run off, and the two materials are separated by gravity. The slag, which is rich in zinc that was present in the ore, is removed to the zinc plant. The molten lead bullion is cooled, the slag, matte, and speiss are removed for metal recovery. The bullion is purified by the addition of sulfur that forms a copper sulfide matte which is sent to a plant for recovery of the copper.

The second step is the refining of the lead produced by the smelting operation. Several processes for refining the bullion are available. In one process, the bullion is charged to a reverberatory furnace in an oxidizing atmosphere, and arsenic, antimony, and tin are removed. Zinc and silver are recovered by other techniques. In another process lead is refined electrolytically or by using molten sodium hydroxide and sodium nitrate as a substitute for furnace softening. Silver and gold are recovered from the zinc slag.

Industrial hygiene hazards in lead smelting are similar to those encountered in copper smelting. The crushing and grinding of ore creates a dust hazard that must be controlled by local exhaust ventilation and the use of water for dust suppression. These operations can also constitute severe noise hazards. On the other hand, the concentration by flotation process creates little dust exposure, and few workers are involved.

Sintering at high temperature releases the sulfur as sulfur dioxide and results in an exposure in the workplace. The hazards in blast furnace operations are exposure to metal dust, fume, carbon monoxide, sulfur dioxide, and heat stress. Arsenic is also released in this operation.

REFERENCES

1 K. W. Nelson, M. Varner, and T. J. Smith, "Nonferrous Metallurgical Operations," in A. Stern, Ed., *Air Pollution,* 3rd ed., Vol. 4, Academic, New York, 1977.

2 W. Wagner, "Environmental Conditions in U.S. Copper Smelters," U.S. Department of Health, Education and Welfare, Publication No. (NIOSH) 75-108, Cincinnati, OH, 1975.

3 "Control Techniques for Lead Air Emissions, Vol. 2, Chap. 4, Appendix B," Publication EPA-450/2-77-012, U.S. Environmental Protection Agency, Research Triangle, NC, 1977.

3.31 STONE QUARRYING

Quarrying refers to the open pit removal of mineral products. The well-known operations involve the bulk removal of such common materials as limestone, clay, and gravel, and dimensional stone products such as granite and marble. Bulk material can be removed from the quarry by conventional earthmoving equipment without major occupational health hazards. Operation of open cab equipment may result in serious exposure to noise from the machinery and to fugitive dust from quarrying operations (1).

The principal hazards occur during quarrying operations that require preparatory drilling for blasting. Rock drilling is done by airpowered percussion drills that deliver power to the cutting bit through a combination of rotation, impact, and thrust (2). If deep holes are required, steel rods are used to connect the power conversion unit, or drifter, to the bit. Compressed air causes the drill mechanism to rotate while rapid blows are delivered by a piston to a striker bar and transmitted through the "string" of drill steels to the bit. Thrust is provided by a separate air motor through a chain drive to the drifter or by hand in the case of hand-held drills such as jack hammers. Cuttings are blown from the hole by blow air introduced through the center of the drill steel and bit.

Sinker drills, or jack hammers, are designed to drill holes from 1.3–6.4 cm (½–2½ in.) in diameter and are primarily used for secondary drilling in breaking up boulders, general excavation, and other jobs where short, small diameter holes are required. Jack hammers will pulverize up to twenty cubic inches of rock per minute which is cleared from the hole by the blow air.

Track drills consist of larger drills mounted on self-propelled crawler carriers that can be driven over very rugged terrain. Most blast hole drilling is done with these machines. Power is supplied through a flexible hose by an air compressor that can be towed by the track drill. The typical drill rig requires approximately 0.28 m³/s (600 cfm) at 620 Pa (90 psig) distributed as follows: blow air, 0.05 m³/s (100 cfm); feed or thrust air, 0.01 m³/s (25 cfm); rotation air, 0.06 m³/s (125 cfm); and percussion or drill air, 0.18 m³/s (350 cfm). Most track drills are designed to drill holes up to about 11.4 cm (4 1–2 in.) in diameter and larger equipment is available for holes up to 15.2 cm (6 in.) in diameter. These drills pulverize 800 to 4000 cm³ (50–250 in.³) of rock per minute.

Considerable work has been done on controlling dust in rock drilling. The most common dust control technique for track drills is water-detergent mist injection. In this method, water, either alone or with a wetting agent such as a detergent, is introduced as a mist into the blow air. The dust collected by the water mist forms small damp pellets that drop out at the edge of the hole. If too much water is used, the pellets form a mud "collar" or a bridge between the steel and the sides of the hole. On the other hand, too little water is ineffective for dust suppression.

The water injection system has no parts exposed to dust abrasion and its efficiency is constant throughout the life of the system. It is also more economical than other control techniques. However, not all the dust is trapped and there are other disadvantages as well. Track drills are frequently operated far from any source of water thus water supply can be a problem. Water tends to displace the lubricant in the system causing increased wear and maintenance costs. The drilling rate is usually reduced with water injection. The slurry tends to harden and form "collars" or irregularities in the hole behind the bit. These have to be broken before the bit will come out, sometimes damaging or losing the steel or bit in the process. In northern latitudes antifreeze must be used to keep the system from freezing. These problems add up to significantly increased operating costs for extra rigs and operators to maintain production.

Mechanical collectors using a reverse air flow through the drill steel or an exhaust hood around the drill steel at the hole collar and powered by air ejectors or mechanical blowers have been developed. These systems are usually made for small drills or very large drills and to date have had limited application on track drills. Dry collectors require less attention than wet methods and have a number of advantages over wet drilling: penetration rate and bit life are better; air alone is more effective in cleaning and stemming blast holes; and the mechanical problems of freezing and transport of water are avoided. The installation is more expensive, however, and internal parts are subject to wear from abrasion leading to decreased efficiency and higher maintenance costs. Low air volume–high velocity dust capture systems offered by certain manufacturers offer a second approach to dust control on penumatic rock drills.

Although several published studies based on midget impinger sampling have shown that dust generated by pneumatic drills can be controlled by water based on midget impinger sampling, a study in our laboratory on track drills under field operating conditions did not show significant reductions in operator exposure when measured with a respirable mass sampler (3). If one is using water for control of dust in quarry drilling operations, one should not assume that control is in place; rather, this hypothesis should be tested by air sampling.

After drilling the blasting operations are commonly done with ammonium nitrate-fuel oil (an-fo) and do not result in a significant exposure in open pits.

In the early quarrying of dimensional stone, the material was removed solely by drilling and blasting operations. This was quite wasteful of stock, and a variety of new quarrying techniques have been developed, some of which present occupational health hazards. The degree of hazard in quarrying dimensional stone depends on the characteristics of the parent rock and, most importantly, its free silica content. In the early 1900s, core drilling was used to block out an island of granite for subsequent removal. This procedure has been replaced by flame cutting using a fuel oil-air burner that cuts through the stone by sloughing it off. This process results in a serious noise hazard and exposure to a particulate cloud consisting of a crystalline respirable dust, a submicrometer rock fume, and a fused micrometer-sized aerosol, as shown in Figure 3.31-1 (4). After this cut has been made, the island of stone is cut into slabs or blocks using a convoluted wire saw that carries an abrasive such as silicon carbide to the cut, as shown in Figure 3.31-2. This operation does not present a health hazard, although erecting the cutting towers can be dangerous. In-quarry cutting provides a semifinished product that can be removed from the quarry using lifting eyes inserted into holes drilled in the individual slabs or blocks. If this drilling is done dry, a significant dust exposure may occur.

The dimensional stone is usually processed at a mill located close to the

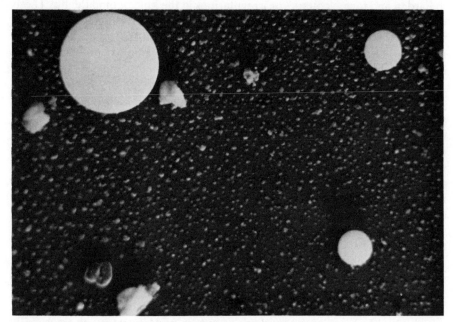

Figure 3.31-1 Particles from burning operation. (From Reference 4, Reprinted courtesy of American Industrial Hygiene Association Journal)

quarry. The production techniques used for architectural products include most of the common hazards. In small job shop operations, the quarry stone is cut to shape by a circular or pit saw using steel shot as an abrasive. Dimensional cuts can be made by diamond saws. Small wire saws are also used for dimensional cutting. Until recently, granite road curbing was cut by first drilling with gang drills, then splitting the stone with wedges. Now, high production of such items as curbstone is done by guillotine splitters. All these operations may be dusty, although the saw cutting, wire saw, and gang drilling usually can be controlled by wet methods. Guillotine cutting is done dry, and the work area must be equipped with local exhaust ventilation.

The finishing operation in the mills includes surfacing with pneumatic tools, small fuel oil-oxygen burners, and polishing tables using various wet abrasives. Sculpting is done by burners, pneumatic tools, and abrasive blasting. The hazard varies with the quartz content of the parent rock. Most of these finishing operations require excellent local ventilation control except polishing. Flame surfacing has not been adequately studied and warrants monitoring.

Stone quarries have a crushing plant for preparation of the final product in the case of a bulk quarry or the recovery of scrap in a dimensional stone quarry. The crushers, transfer points, and screening operations all require ventilation

Figure 3.31-2 Quarry operation.

control. The dustiest operation is usually bagging, however, modern bagging equipment now available includes integral exhaust hooding.

The present ventilation standards for granular material transfer points are deficient; however a proposed induced-flow calculation method provides improved control (5).

In evaluating exposure in quarries, it is extremely important that personal monitoring be conducted; personal experience has shown little relationship between fixed location and personal monitoring. In siliceous dust such as that encountered in granite and slate quarries, the silica exposure should be measured by respirable mass determinations on personal air samples. It is not uncommon for the parent rock to contain 30% free silica and the respirable sample only 10% free silica.

REFERENCES

1 A. F. Burns, F. Ottoboni, and H. Mitchell, *Am. Ind. Hyg. Assoc. J.*, **23,** 273 (1962)

2 The material presented on rock drilling was contributed by Ray Woodcock.

3 R. Woodcock and W. Burgess, Observations on Dust Control on Track Drills, unpublished data, 1977.

4 W. A. Burgess and P. Reist, *Am. Ind. Hyg. Assoc. J.*, **30,** 107 (1969).

5 D. Anderson, *Ind. Med. Surg.*, **133,** 68 (1964)

3.32 UNDERGROUND MINING

Diverse minerals are mined, both underground and in surface or open cast mines. This discussion is limited to underground mining operations. The health hazards include dust exposures to the mineral being recovered and its associated rock, natural and man-made gases and vapors, and a gamut of physical hazards. The subject is too great to be covered in depth; however, various common occupational problems arise in underground mining operations.

A principal problem usually is associated with the mineral being extracted or the materials associated with the ore body. In the case of asbestos and mercury, the recovered material may present the major hazard, whereas the high concentrations of quartz in hard rock mining for lead and zinc ore present the major dust hazard in those mines. One should know the complete geological characteristic of the ore body, not merely the main mineral constituents, to be able to evaluate this dust exposure properly.

Coal mining presents special dust hazards when high-energy mining methods are employed. The conventional mining procedures of cutting, drilling, mining, and loading the coal generally have been replaced by more energy intensive, high production systems. A common coal mining method in the United States is the use of a continuous miner, which accomplishes all the foregoing operations with a single piece of equipment. Long wall mining is more popular in the United Kingdom than in the United States because of differences in coal seam geometry. These mining methods generate considerable dust and warrant special control techniques (1). In the open hearings on the Coal Mine Safety and Health Act of 1969, it was claimed that dust control technology in coal mines would not be adequate to achieve the proposed standard at 5 mg/m^3. However, tremendous strides in dust control in coal mines have been made, and most working faces now meet the early standard.

A number of exposures to mine gases and vapors are listed in Table 3.32-1. The natural gaseous emissions in coal mines and certain hard rock mines are methane, carbon dioxide, and nitrogen. Because of its wide flammability range (5 to 15%), methane presents a major hazard. Firedamp explosions in coal mines can initiate subsequent violent dust explosions, accompanied by the release of carbon monoxide.

Table 3.32-1 *Common Names of Mine Gases*

Type of Gas	Common Name
Methane	Firedamp
Carbon monoxide	Whitedamp
Hydrogen sulfide	Stinkdamp
Oxygen deficiency	Blackdamp
Gases from explosives	Afterdamp

In addition to natural sources, man's underground activity generates other toxic gases and vapors. Straight dynamite (100% nitroglycerin) is not employed underground, but a blend of dynamite and ammonium nitrate or other explosives, compounded to minimize the release of toxic gases, is in common use. Recent practice is to use ammonium nitrate prills saturated with fuel oil and fired with dynamite caps (2). The principal gaseous contaminants continue to be carbon monoxide and nitrogen dioxide.

The widespread use of diesel engines underground contributes exhaust gases containing aldeyhdes, nitrogen dioxide, and carbon monoxide (3). Scrubbers are used frequently on such equipment to remove aldehydes. Urethane foam systems for sealing brattice and leaks have been introduced underground, and this practice results in exposure to diphenyl methane diisocyanate.

Oxygen deficiency, especially in old workings when a reducing ore exists or organic material is decomposing, can be a major hazard. Significant radon daughter concentrations occur not only in uranium mines but in other underground works, and excellent ventilation controls are required in such cases. In mineral mines high radon concentrations are encountered in areas that have not been worked recently. The physical hazards in mining include temperature and humidity extremes, poor lighting, and noise and vibration. The rock temperatures underground increase at a rate of 1°C for each 100 m of depth. The high rock temperature in deep mines and the extensive use of water for dust suppression may cause a serious heat stress problem. It is not uncommon to place air-conditioning plants underground to cool the air delivered to working faces. Personal cooling systems have also been developed.

The widespread use of percussion drills and other compressed air tools presents a serious noise hazard to the underground miner. This problem can be controlled fully only by equipment redesign, prompted by rigorous purchasing specifications. Serious loss of hearing has been noted in mining populations. A study of vibration disease in this population has not been attempted to date.

Certain necessary controls of occupational health hazards have been cited. The foremost hazard, dust exposure to the toxic mineral dust or silica exposure from the host rock, calls for special attention. Wet methods are a principal control, and water is used with percussion drills, to infuse working faces at the cutting head of continuous miners, and to wet down loose rock. Wetting agents are used to improve the effectiveness of the wet procedures. Comments are made on the effectiveness of wet drilling in Section 3.31. To minimize dust explosions from settled coal dust, an inert dust such as limestone is employed to dust all surfaces. Equipment redesign, especially of cutting tools, can be helpful in reducing dustiness. The effects of dust from blasting is minimized by conducting this operation when the men are not in the area. Wetting down with water before and after blasting is important also.

The principal control is, of course, ventilation. Occasionally local exhaust with air cleaning and direct return of cleaned air can be accomplished underground at specific dusty locations, such as crushers and conveyor transfer points. For the most part, however, dilution ventilation is necessary for both methane and dust control.

REFERENCES

1 K. M. Morse, *Am. Ind. Hyg. Assoc. J.*, **31,** 160 (1970).
2 "Michigan's Metal Mines Underground Health Hazards," *Mich. Occup. Health,* **12,** 4 (Spring 1967).
3 "The Use of Diesel Equipment In Underground Coal Mines," Reports from Workshops held at Morgantown, WV, Sept. 19–23, 1977, U.S. Department of Health, Education and Welfare, NIOSH, Morgantown, W.VA, Feb. 1978.

COMPOSITION OF METAL ALLOYS

Several organizations have devised identification systems for metals that are useful to occupational health specialists. The most frequently used systems are those proposed by the American Iron and Steel Institute (AISI); however, other systems are published by the American Society for Metals (ASM), Society of Automotive Engineers (SAE), American Society for Testing and Materials (ASTM), and the Steel Founders Society of America (SFSA). Examples of the composition of alloy steels, stainless steels, nickel alloys, casting alloys, and tool steels are shown in Tables A-1 to A-5.

Table A-1 *AISI Designation System for Alloy Steels*

Alloy Series	Approximate Alloy Content (%)
13XX	Mn 1.60–1.90
40XX	Mo 0.15–0.30
41XX	Cr 0.40–1.10; Mo 0.08–0.35
43XX	Ni 1.65–2.00; Cr 0.40–0.90; Mo 0.20–0.30
44XX	Mo 0.45–0.60
46XX	Ni 0.70–2.00; Mo 0.15–0.30
47XX	Ni 0.90–1.20; Cr 0.35–0.55; Mo 0.15–0.40
48XX	Ni 3.25–3.75; Mo 0.20–0.30
50XX	Cr 0.30–0.50
51XX	Cr 0.70–1.15
E51100	C 1.00; Cr 0.90–1.15
E52100	C 1.00; Cr 0.90–1.15
61XX	Cr 0.50–1.10; V 0.10–0.15

Table A-1 *AISI Designation System for Alloy Steels—Continued*

Alloy Series	Approximate Alloy Content (%)
86XX	Ni 0.40–0.70; Cr 0.40–0.60; Mo 0.15–0.25
87XX	Ni 0.40–0.70; Cr 0.40–0.60; Mo 0.20–0.30
88XX	Ni 0.40–0.70; Cr 0.40–0.60; Mo 0.30–0.40
92XX	Si 1.80–2.20

Source. From *Machine Design, 1971 Reference Issue,* Copyright 1971, Used with permission of Machine Design.

Table A-2 *AISI Stainless Steel Classification System*

AISI No.	Chemical Analyses of Stainless Steels (%)					
	Carbon	Manganese	Silicon	Chromium	Nickel	Other Elements
Chromium-Nickel-Magnesium-Austenitic-Nonhardenable						
201	0.15 Max.	5.5/7.5	1.0	16.0/18.0	3.5/5.5	N_2 0.25 Max.
202	0.15 Max.	7.5/10.	1.0	17.0/19.0	4.0/6.0	N_2 0.25 Max.
Chromium-Nickel-Austenitic-Nonhardenable						
301	0.15 Max.	2.0	1.0	16.0/18.0	6.0/8.0	—
308	0.08 Max.	2.0	1.0	19.0/21.0	10.0/12.0	—
309	0.20 Max.	2.0	1.0	22.0/24.0	12.0/15.0	—
Chromium-Martensitic-Hardenable						
403	0.15 Max.	1.0	0.5	11.5/13.0	—	—
410	0.15 Max.	1.0	1.0	11.5/13.5	—	—
Chromium-Ferritic-Nonhardenable						
405	0.08 Max.	1.0	1.0	11.5/14.5	—	A1 1.1/0.3
430	0.12 Max.	1.0	1.0	14.0/18.0	—	—
430F	0.12 Max.	1.25	1.0	14.0/18.0	—	S 0.15 Min.
430FSe	0.12 Max.	1.25	1.0	14.0/18.0	—	Se 0.15 Min.

Table A-3 *Composition of Nickel and Nickel Alloys*

Alloy Designation	Nominal Chemical Composition (%)										
	Ni	C	Mn	Fe	S	Si	Cu	Cr	Al	Ti	Cb
Nickel 200	99.5	0.08	0.18	0.2	0.005	0.18	0.13	—	—	—	—
Nickel 201	99.5	0.01	0.18	0.2	0.005	0.18	0.13	—	—	—	—
Monel alloy 400	66.5	0.15	1.0	1.25	0.012	0.25	31.5	—	—	—	—
Monel alloy 401	42.5	0.05	1.6	0.38	0.008	0.13	Bal.	—	—	—	—

Table A-3 *Composition of Nickel and Nickel Alloys—Continued*

Alloy Designation	Nominal Chemical Composition (%)										
	Ni	C	Mn	Fe	S	Si	Cu	Cr	Al	Ti	Cb
Inconel alloy 600	76.0	0.08	0.5	8.0	0.008	0.25	0.25	15.5	—	—	—
Inconel alloy 601	60.5	0.05	0.5	14.1	0.007	0.25	0.50	23.0	1.35	—	—
Incoloy alloy 800	32.5	0.05	0.75	46.0	0.008	0.50	0.38	21.0	0.38	0.38	—
Incoloy alloy 801	32.0	0.05	0.75	44.5	0.008	0.50	0.25	20.5	—	1.13	—

Table A-4 *Chemical Compositions of Tool Steels*

	AISI Designation	C (%)	Distinguishing Alloying Elements							
			Mn	Si	Ni	Cr	Mo	W	V	Co
Water-hardening	W	0.60–1.40				X			X	
Shock-resisting	S	0.50–0.55	X	X		X	X	X		
Oil-hardening	O	0.90–1.45	X	X		X	X	X		
Air-hardening	A	0.50–2.25	X	X	X	O	O	X	X	
Cold-work	D	1.50–2.35				O	O		X	X
Hot-work (chromium)	H	0.35–0.40				O	X	X	X	X
Hot-work (tungsten)	H	0.25–0.50				O		O	X	
Hot-work (molybdenum)	H	0.55–0.65				O	O	X	O	
High-speed (tungsten)	T	0.75–1.50				O		O	O	X
High-speed (molybdenum)	M	0.80–1.30				O	O	O	O	X
Low-alloy	L	0.50–1.10			X	O			X	
Carbon-tungsten	F	1.00–1.25						O		
Mold and die	P	0.07–0.35			X	X	X			

Source. From *Machine Design, 1971 Reference Issue,* Copyright 1971, Used with permission of Machine Design.

O = most tool steels of this type contain significant amounts of this element.
X = several—but not all—tool steels of this type contain this element.

Table A-5 *Composition of Selected Casting Alloys*

Alloy		
Name	ASTM No.	Typical Composition (wt. %)
Ferrous		
Cast steel	60-30	≤0.25C
	175-145	
Gray iron	20	3.5C, 2.4Si, 0.4P, 0.1S
	60	2.7C, 2.0Si, 0.1P, 0.1S, 0.8Mn
Malleable iron	A47	2.5C, 1.4Si, 0.1P, 0.1S, 0.4Mn
Nodular iron	60-40-15	3.5C, 2.4Si, 0.1P, 0.03S, 0.8Mn
Stainless steel	CF8	0.08C, 19Cr, 9Ni
Cu-base		
Leaded red brass	4A	5Sn, 5Pb, 5Zn
High-lead tin brass	3A	10Sn, 10Pb
Leaded yellow brass	6C	1Sn, 1Pb, 37Zn
Al-base		
	108	4Cu, 3Si
	D132	3.5Cu, 9Si, 0.8Mg, 0.8Ni
	380	3.5Cu, 8Si
Mg-base		
	AZ91	9A1, 0.7Zn, 0.2Mn
	EZ33A	2.7Zn, 0.5Zr, 3 rare earths
Zn-base		
	AG40A	4A1, 0.04Mg
Pb-base		
Type metal		3Sn, 11Sb
Sn-base		
Babbitt	Alloy 1	4.5Sb, 4.5Cu

Source: From *Introduction To Manufacturing Processes.* Copyright 1977. Used with permission of McGraw-Hill Book Co.

SOURCES OF INFORMATION

1 NATIONAL ORGANIZATIONS

Air Pollution Control Association.   4450 Fifth Avenue, Pittsburgh, PA 15213

The American Academy of Industrial Hygiene.   475 Wolf Ledges Parkway, Akron, OH 44311.

The American Academy of Occupational Medicine.   150 North Wacker Drive, Chicago, IL 60606.

American Association of Occupational Health Nurses, Inc.   (AAOHN).   79 Madison Avenue, New York, NY 10016.

American Board for Occupational Health Nurses, Inc.   (ABOHN).   521 W. Westfield Avenue, Roselle Park, NJ 07204.

American Chemical Society.   1155 Sixteenth Street, N.W., Washington, D.C. 20036

American Conference of Governmental Industrial Hygienists, Inc.   Executive Secretary, Mr. W. D. Kelley. 2205 South Road, Cincinnati, OH 45238.

American Industrial Hygiene Association.   William E. McCormick, Managing Director. 475 Wolf Ledges Parkway, Akron, OH 44311.

American Medical Association.   535 North Dearborn Street, Chicago, IL 60610.

American National Standards Institute.   1430 Broadway, New York, NY 10018.

American Occupational Medical Association.   150 North Wacker Drive, Chicago, IL 60606.

American Public Health Association.   1015 Eighteenth Street, N.W., Washington, D.C. 20036.

American Society for Testing and Materials.   1916 Race Street, Philadelphia, PA 19103.

The British Occupational Hygiene Society.   Hon. Secretary, Mr. J. T. Sanderson. Esso Research Centre, Abingdon, Oxon OX13 6AE, England.

Chemical Industry Institute of Toxicology.   Research Triangle Park, NC 27709.

Industrial Health Foundation, Inc.   5231 Centre Avenue, Pittsburgh, PA 15232.

National Safety Council.   425 N. Michigan Avenue, Chicago, IL 60611.

Society of Toxicology.   475 Wolf Ledges Parkway, Akron, OH 44311.

2   INDUSTRIAL AND TRADE ASSOCIATIONS

American Iron and Steel Institute.   1000 16th Street, N.W., Washington, D.C. 20036.

The Aluminum Association.   818 Connecticut Avenue, N.W., Washington, D.C. 20006.

American Foundrymen's Society.   Golf and Wolf Roads, Des Plaines, IL 60016.

American Petroleum Institute.   2101 L Street, N.W., Washington, D.C. 20037.

American Society for Metals.   Metals Park, OH 44073.

American Welding Society.   2501 Northwest Seventh Street, Miami, FL 33125.

Chemical Manufacturers' Association.   1825 Connecticut Avenue, N.W., Washington, D.C. 20009.

Compressed Gas Association, Inc.   500 Fifth Avenue, New York, NY 10036.

The Fertilizer Institute.   1015 18th Street, N.W., Washington, D.C. 20036.

Forging Industry Association.   1121 Illuminating Building, Cleveland, OH 44113.

Investment Casting Institute.   8521 Clover Meadow, Dallas, TX 75243.

Metal Finishers Supplies Association.   1025 E. Menle Road, Birmingham, MI 48011.

Metal Powder Industries Federation.   P. O. Box 2054, Princeton, NJ 08540.

Metal Treating Institute.   1300 Executive Center, Suite 115, Tallahassee, FL 32301.

National Association of Metal Finishers.   111 E. Wacker Drive, Chicago, IL 60601.

Society of Manufacturing Engineers.   P. O. Box 930, Dearborn, MI 48128.

The Society of the Plastics Industry.   355 Lexington Avenue, New York, NY 10017.

3 FEDERAL AGENCIES

Department of Commerce

National Bureau of Standards. Washington, D.C. 20234.

National Technical Information Service (NTIS). 5285 Port Royal Road, Springfield, VA 22151.

Department of Health and Human Services

Public Health Service

Center for Disease Control. Atlanta, GA 30333.

National Institute for Occupational Safety and Health (NIOSH). Parklawn Building, 5600 Fishers Lane, Rockville, MD 20852.

Bureau of Radiological Health. 12720 Twinbrook Parkway, Rockville, MD 20852.

National Clearinghouse for Poison Control Centers. 5401 Westbard Avenue, Bethesda, MD 20016.

National Institutes of Health

National Cancer Institute. Bethesda, MD 20014.

National Heart and Lung Institute. Bethesda, MD 20014.

National Institute of Environmental Health Sciences (NIEHS). P. O. Box 12233, Research Triangle Park, NC 27709.

National Library of Medicine. 8600 Rockville Pike, Bethesda, MD 20014.

Department of the Interior

Mine Safety and Health Administration. C Street between Eighteenth and Nineteenth Streets, Washington, D.C. 20240.

Department of Labor

Bureau of Labor Statistics. 200 Constitution Avenue, N.W., Washington, D.C. 20210.

Occupational Safety and Health Administration. 2000 Constitution Avenue, Washington, D.C. 20210.

Environmental Protection Agency

Air and Water Programs Office. 401 M Street, S.W., Washington, D.C. 20460.

Pesticides Programs Office. 401 M Street, S.W., Washington, D.C. 20460.

Radiation Programs Office. Rockville, MD 20852

Water Program Operations Office. 401 M Street, S. W., Washington, D.C. 20460.

Air Pollution Technical Information Center (APTIC), Office of Technical Information and Publications (OTIP), Air Pollution Control Office (APCO). P. O. Box 12055, Research Triangle Park, NC 27709.

National Environmental Research Center. Research Triangle Park, NC 27711 (also located in Cincinnati, OH 45268; Corvallis, OR 97330; and Las Vegas, NV 89114).

Western Environmental Research Laboratory. P. O. Box 15027, Las Vegas, NV 89114.

National Council for Radiation Protection and Measurements (NCRP). 7610 Woodmont Avenue, S.E., 1016, Washington, D. C. 20014.

Nuclear Regulatory Commission. 1717 H Street, N.W., Washington, D. C. 20555. U.S. Government Printing Office. Washington, D.C. 20402.

4 UNIONS

AFL-CIO Dept. of Health and Safety
AFL-CIO Building, 815 16th St., N.W.
Room 507
Washington, D.C. 20006

Amalgamated Clothing & Textile Workers Union
770 Broadway
New York, New York 10003

Boilermakers, International Brotherhood of—
    Local 802
P.O. Box 618
Chester, Pennsylvania 19016

Chemical Workers Union, International
1655 W. Market Street
Akron, Ohio 44313

IUE—Local 201
Health and Safety Committee
100 Bennett Street
Lynn, Massachusetts 01905

Molders and Allied Workers Union
1216 E. McMillan Street
Suite 302
Cincinnati, Ohio 45206

Oil, Chemical and Atomic Workers
1636 Champa
P.O. Box 2812
Denver, Colorado 80201

Rubber, Cork, Linoleum and Plastic Workers
  of America, United
URWA Building
87 S. High Street
Akron, Ohio 44308

Steelworkers of America, United
Five Gateway Center
Pittsburgh, Pennsylvania 15222

United Auto Workers
Solidarity House
8000 East Jefferson Avenue
Detroit, Michigan 48214

Teamsters, International Brotherhood of
25 Louisiana Avenue, N.W.
Washington, D.C. 20001

5   REFERENCES ON ODOR THRESHOLDS

G. Leonardos, D. Kendall, and N. Barnard, *J. Air Pollut. Control Assoc.*, **19** (2), 91–95 (1969).

F. V. Wilby, *J. Air Pollut. Control Assoc.*, **19**, 96–100 (1969).

T. M. Hellman and F. H. Small, *J. Air Pollut. Control Assoc.*, **24** 979–982 (1974).

F. Patte, M. Etcheto, and P. Laffort, *Chem. Senses Flavor,* **1**, 283–305 (1975).

Anton Naus, *Olphactoric Properties of Industrial Matters*, Charles University, Prague, 1976.

Von Eberhard Hill, *Staub-Reinhalt Luft*, **37** (5), 199–201 (1977).

6   REFERENCES ON MATERIALS AND PROCESSES

M. Begeman and B. Amstead, *Manufacturing Processes*, Wiley, New York, 1969.

W. J. Patton, *Modern Manufacturing Process and Engineering*, Prentice-Hall, Englewood Cliffs, NJ, 1970.

C. R. Rhine, *Machine Tools and Processes for Engineers*, McGraw-Hill, New York, 1971.

V. B. John, *Introduction to Engineering Materials*, Macmillan, London, 1972.

L. H. Van Vlack, *A Textbook of Materials Technology*, Addison-Wesley, Reading, MA 1973.

Metals Handbook, 8th ed., American Society for Metals, Metals Park, OH.

  Vol. 3, *Machining*, 1967.

  Vol. 5, *Forging and Casting*, 1970.

  Vol. 6, *Welding and Brazing*, 1971.

Modern Plastics Encyclopedia, Modern Plastics, Highstown, NJ, (annual).

Plastics Engineering Handbook, The Society of the Plastics Industry, Inc., 1960.

J. A. Kent, Ed., *Riegel's Handbook of Industrial Chemistry*, 7th ed., Van Nostrand Reinhold, New York, 1974.

D. M. Considine, Ed., *Chemical and Process Technology Encyclopedia,* McGraw-Hill, New York, 1974.

R. H. Perry, Ed., *Chemical Engineers Handbook*, 5th ed., McGraw-Hill, New York, 1973.

Parts 1 and 3 of Appendix B are based on Section XI *Occupational Diseases, A Guide to Their Recognition*, rev. ed., DHEW Publication No. (NIOSH) 77-181, Washington, D.C. 1977.

NOMENCLATURE

1 FORM OF AIR CONTAMINANTS

1.1 *Particulates*

1.1.1 Dust—formed from solid material by mechanical processes such as grinding, blasting and drilling. Examples: silica, talc, asbestos, and lead dusts.

1.1.2 Fume—small, solid particles formed by vaporization and subsequent condensation. Example: cadmium fume produced by heating the metal to form submicrometer particulate that oxidizes to cadmium oxide.

1.1.3 Mist—liquid particulates formed by direct atomization from a liquid or condensation from the gaseous state. Example: oil mist from coolants on machinery operations, paint mist from spray painting, and plating mists.

1.2 *Gases*—formless fluids that occupy the entire space of the enclosure and can be changed to a liquid or solid state only under increased pressure or decreased temperature. Examples: hydrogen sulfide, carbon monoxide, and chlorine.

1.3 *Vapors*—evaporation products of substances normally liquid or solid at normal temperature and pressure. Examples: carbon disulfide and trichloroethylene.

2 ACRONYMS FOR ORGANIZATIONS

AAIH   American Academy of Industrial Hygiene
ABIH   American Board of Industrial Hygiene
ACGIH   American Conference of Governmental Industrial Hygiene

ACIL   American Council of Independent Laboratories
ACS   American Chemical Society
AIChE   American Institute of Chemical Engineers
AIHA   American Industrial Hygiene Association
AIME   American Institute of Mining, Metallurgical, & Petroleum
                Engineers
AMA   American Medical Association
ANS   American Nuclear Society, Inc.
ANSI   American National Standards Institute
APCA   Air Pollution Control Association
APHA   American Public Health Association
ASA   Acoustical Society of America
ASHRAE   American Society of Heating, Refrigerating, and Air Condi-
                tioning Engineers
ASSE   American Society of Safety Engineers
ASTM   American Society for Testing and Materials
BCSP   Board of Certified Safety Professionals of the Americas, Inc.
BOHS   British Occupational Hygiene Society
CFR   Code of Federal Regulations
CMA   Chemical Manufacturers Association
CSA   Construction Safety Act
                (Contract Work Hours and Safety Standards Act)
CSHO   Compliance Safety and Health Officer, OSHA
DOD   Department of Defense
DOL   Department of Labor
DOT   Department of Transportation
EPA   Environmental Protection Agency
FM   Factory Mutual System
HFS   Human Factors Society, Inc.
HHS   Department of Health and Human Services
ICC   Interstate Commerce Commission
IHF   Industrial Health Foundation of America, Inc.
IMA   Industrial Medical Association, Inc.
ISEA   Industrial Safety Equipment Association, Inc.
NACOSH   National Advisory Committee on Occupational Safety and
                Health
NBS   National Bureau of Standards, Department of Commerce
NECA   National Electrical Contractors Association
NFPA   National Fire Protection Association

NIOSH	National Institute for Occupational Safety and Health
NSC	National Safety Council
OSHA	Occupational Safety and Health Administration, U.S. Department of Labor
OSHRC	Occupational Safety and Health Review Commission
PCA	Public Contracts Act (Walsh-Healey)
SCA	Service Contracts Act
SIC	Standard Industrial Classification
UL	Underwriters Laboratories, Inc.

Source. National Institute For Occupational Safety and Health.

INDEX